MW00641487

Forty Ways to Know a Star

Forty Ways to Know a Star

USING STARS TO UNDERSTAND ASTRONOMY

Jillian Scudder, PhD

PA PRESS

PRINCETON ARCHITECTURAL PRESS · NEW YORK

✴ Know a star. . .

introduction

Looking out at the night sky, the stars may appear as nothing more than faint pinpricks of light interrupting the darkness. When observed more carefully, differences begin to jump out, even to the unaided eye. We can find stars that are redder than others, stars that are brighter, and some that are more densely collected together. If we call on assistance from humanity's many telescopes, these differences become even more stark.

Beginning with these differences in color and brightness, and aided by our ability to observe our very own star—the Sun—close up and carefully, we've built up an understanding of how the stars function. We now know where they form their energy, how that energy can escape their depths, and what their outermost layers look like. This knowledge means that the impacts of living near our own star go far beyond simply the warmth needed to host life.

The explosive and destructive ends of the lives of stars have been observed both by ancient civilizations and modern technologies. Through state-of-the-art observations and models of these cataclysms, we've learned how to understand what remains after these detonations, if anything, how the elements we see on Earth came to be, and the particular debt Earth's gold and silver owes to the end of a star's existence.

These foundations have allowed us to understand even more complicated stars. We've learned about stars that change in brightness over time, some of them so predictably that we've been able to use them as distance markers in a Universe that they themselves have proved to be much larger than we had thought. The stars have taught us about invisible components of our Universe, which turn out to vastly outnumber the visible portions, and their glow has taught us about how galaxies formed and changed over cosmic time.

All along humanity's journey to understand the Universe, it is the stars that have led and lit the way. In this book, I introduce forty different ways to understand the stars, the galaxies they inhabit, and the cosmos they illuminate. I hope you enjoy the journey, and come away with a deeper connection to those faint pinpricks of light.

as a light in the sky

To begin to know the stars, we can simply gaze up into the dark night. When the Sun sets, and your eyes adjust to the darkness, your vision can capture light that's been traveling for hundreds to thousands of years to reach the Earth—the light from the stars.

In extremely dark sky conditions, about 9,100 stars are visible to the unassisted human eye, and in moonless, ideal conditions, it's an impressive sight, with brilliant points of light peppering the entire sky above you.

Humans have used the stars for storytelling, navigating, and measuring the seasons as far back as records go. Oral histories confirm that stargazing is a fundamentally human venture. While most of us don't navigate by the stars anymore, storytelling has remained embedded in the names of the constellations. These come to us from the ancient Greeks, though of course every civilization had their own set of constellations and stories. The constellations refer to the characters of many Greek myths: the hero Perseus, whose story involves the beheading of Medusa, the winged horse Pegasus, and the rescue of Andromeda from a sea monster, Cetus. Medusa is represented as the star Algol within the constellation of Perseus; Andromeda sits nearby Perseus. Pegasus is visible in the summer sky as the Great Square of Pegasus (it's a big square of bright stars). Cetus has no bright stars in it, but it's still present as a region in the sky (fig. 1.1).

In modern astronomy, the International Astronomical Union (IAU) agreed in 1922 to use a set of eighty-eight constellations covering the entire night sky. The boundaries between those constellations were published in 1930 by Eugène Delporte, with the approval of the IAU. Since then, those boundaries and constellations have provided an easy way to quickly indicate where a star is in the sky—it will always be within some known, standardized constellation.

Most people in the world live alongside city lights and other artificial sources of illumination, and as these lights become brighter, the faintest stars disappear from view. By a 2016 estimate, 14 percent of the global population, 20 percent of the EU, and 37 percent of the

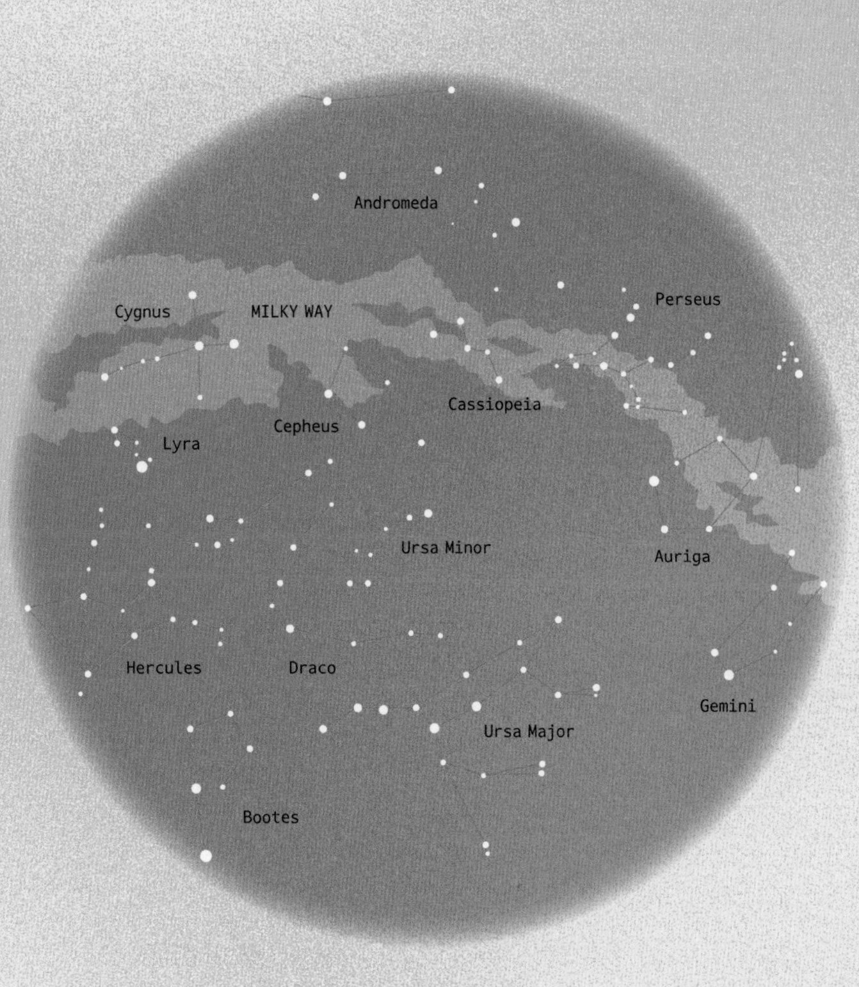

FIG. 1.1	The boundaries of the constellations were finalized only in 1930, but now serve as a first point of reference for finding astronomical objects in the sky.

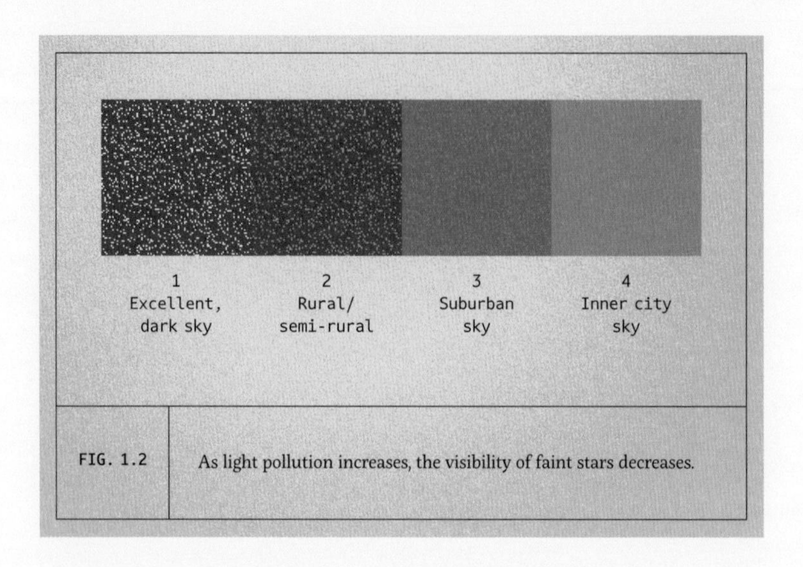

FIG. 1.2 As light pollution increases, the visibility of faint stars decreases.

US population lives in such city-illuminated regions that the skies are never truly dark and the eyeball never fully switches into night-vision mode. Singapore is the most affected—100 percent of the population in Singapore will not see anything darker than twilight in their skies (fig. 1.2).

Unless you live in a major city, you're likely to see somewhere between the 9,100 stars of the darkest skies and those of the most light-polluted cities. But the brighter the star, the rarer they are—there are only forty-five stars in the night sky brighter than Polaris, the north star. The Orion Nebula, by contrast, which masquerades as a star in the sword of Orion, is fainter. If you can see that object, then another 513 stars would be bright enough to be seen (though of course, some of them will be under the horizon).

There's an increasing push to preserve the remaining truly dark places in the world as dark sky sites: locations people could visit to acquaint themselves with what the night sky would have looked like to every human prior to the advent of industrial lighting. They tend to be fairly remote, unfortunately—large cities can scatter their light much further afield than one might expect. In the US, a number of national parks are classified as International Dark Sky Places, and the skies there are usually spectacular.

Stars do appear to twinkle in the sky, especially if it's a windy night. This twinkling isn't something the light from the star is doing as it leaves its star. Instead, it's the impact of the Earth's atmosphere. As you go from sea level up to the upper atmosphere, the temperature of the atmosphere drops dramatically, but if you look carefully, there are little pockets of warmer and cooler air, instead of a perfectly smooth gradient. Each little pocket can bend the incoming beam of light from the star, and so by the time the light makes it down to us, observing from the surface of the Earth, the dot of light can seem to flicker in brightness as our atmosphere blurs it in and out of focus. Since the beam of light from a star is so thin and narrow, even small disturbances become visible to our eyes (fig. 1.3).

The more pockets of air there are, the larger this impact is, and the easiest way to get a lot of small pockets of air is for the atmosphere to be windy. It doesn't have to be windy at ground level, and it often isn't, but if you see the stars twinkling and the air is still where you are, then you can have a pretty good guess that high above you, the atmosphere is blustery.

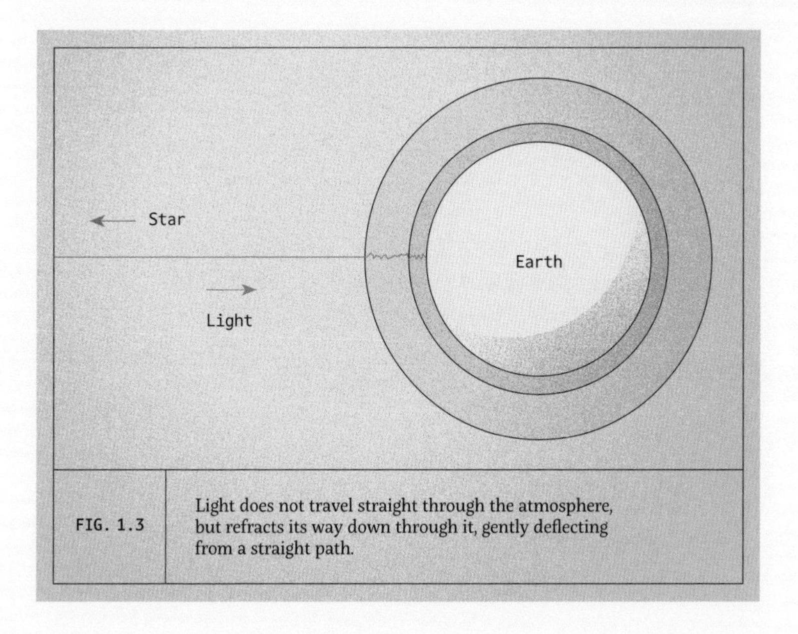

FIG. 1.3 Light does not travel straight through the atmosphere, but refracts its way down through it, gently deflecting from a straight path.

as the source of daylight

Our own daytime hours are an easy way to appreciate the stars. For all the beauty and fantastical diversity of the night sky, there is one star in particular that so vastly outshines the rest of them that when it's above our horizon, it floods the sky with so much light that seeing other stars is functionally impossible: our Sun.

The Sun is not bright in our skies because it's particularly brighter than the rest of the stars, but simply because we're so close to it. "Close" is an astrophysically relative term here—the distance between the Earth and the Sun is still a mind-boggling 93 million miles (150 million kilometers). But compared to the next closest star, Proxima Centauri, which sits 268,770 times further, we're positively cozy with our Sun.

With a specially protected telescope, the Sun is an easy target to observe, so long as it's not cloudy. From the ground, we can already begin to see things beyond a simple bright disk in the sky. Observations of the Sun well predate the telescope, with written records going back to 800 BCE in China. The first known drawing of a sunspot comes from 1128 CE, in John of Worcester's *Chronicle*. Sunspots are dark patches on the Sun, where the surface is a bit cooler than the surrounding area. They arise whenever the magnetic field of the Sun becomes tangled and pops a loop out of the surface.

The tangling of the magnetic field happens because the equator of the Sun rotates faster than the poles—something we learned by watching how fast it takes a sunspot to cross the disk of the Sun. The more times the equator laps the poles, the more complex the magnetic field becomes, the more loops it's likely to create, and the more likely it is for sunspots to appear on its surface (fig. 2.1).

Without technology to help, and under normal circumstances, sunspots are the only major defect in the Sun that can be seen with the human eye. When the magnetic field of the Sun is calm, observed carefully, our star looks like a featureless bright disk. However, with technological advances, we've been able to see the dynamic and ever changing features on the surface of the Sun. Ground-based telescopes have been able to image the Sun in

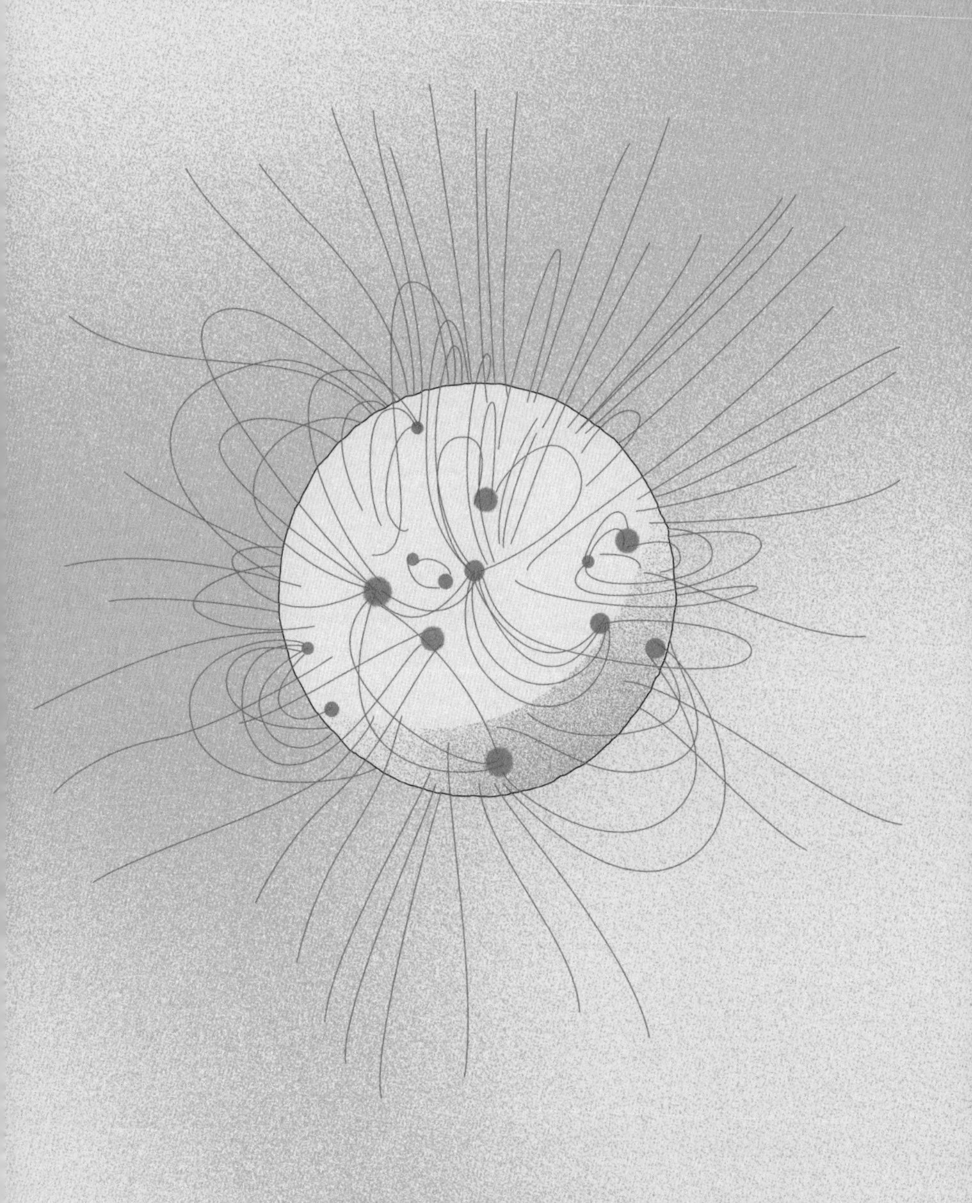

FIG. 2.1	The magnetic field of the Sun is anything but straightforward, and tangles often burst through the surface, leaving the visible trace of a sunspot.

enormous detail, showing a roiling surface with bubbles of intensely heated plasma rising in a constantly shifting glitter known as granulation. These bubbles are bounded by slightly cooler plasma, sinking back down into the depths of the star, in a process most akin to boiling water. The smallest granulation cells are roughly the size of Texas, with larger ones spanning a surface area that would cover the majority of the US.

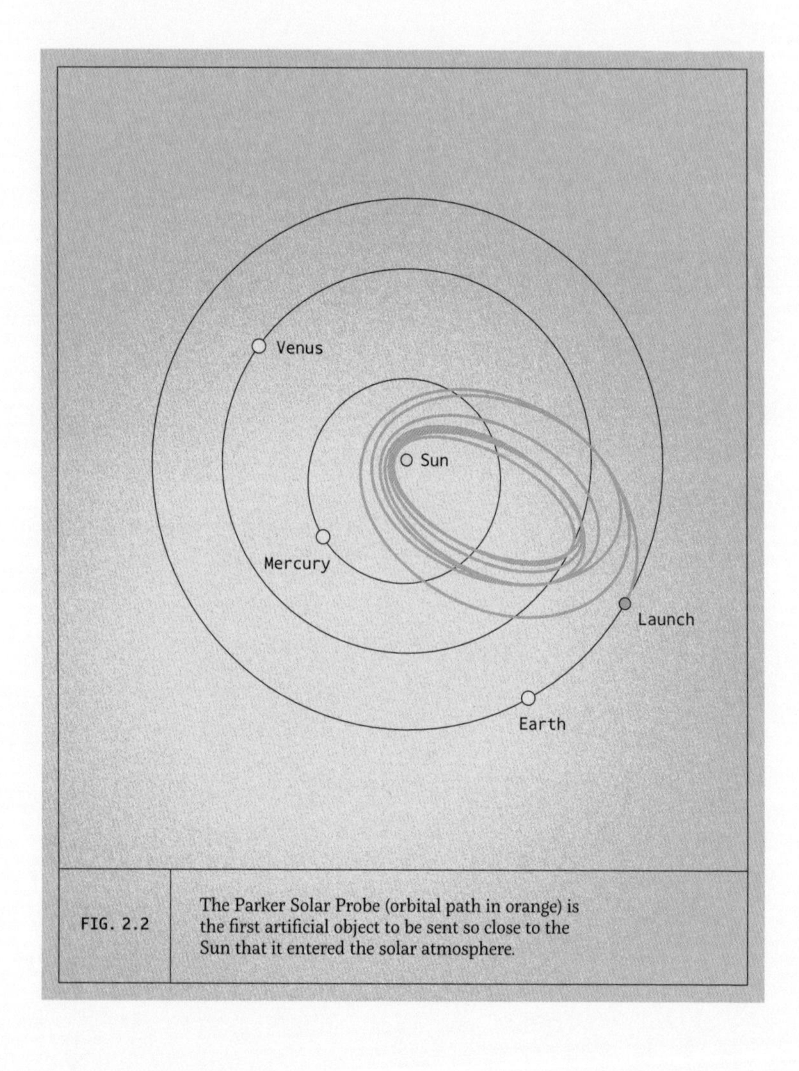

FIG. 2.2 The Parker Solar Probe (orbital path in orange) is the first artificial object to be sent so close to the Sun that it entered the solar atmosphere.

We have also launched telescopes out into space to observe the Sun, without the hindrance of being offline for half of every day. Two of the most productive of these are the Solar Dynamics Observatory (SDO) and the Solar and Heliospheric Observatory (SOHO), both of which have returned a number of impressive images capturing detailed views of the Sun. SOHO has been placed in an orbit with an unobscured view of the Sun, so that the satellite is in permanent daylight. SDO orbits the Earth, but in a wide orbit where its view of the Sun is only occasionally blocked by our planet. SDO and SOHO combine to give us particularly good vistas of any dramatic changes on the solar surface, such as solar prominences. These are large loops of plasma, carried high above the surface along magnetic currents, many tens of times larger than the Earth. A prominence can persist over days to months, and during that time, the plasma caught up in the magnetic loop can gradually rain back down onto the surface of the Sun—this is dubbed coronal rain, and is certainly the least refreshing form of "rain" you'll encounter.

The Parker Solar Probe, as of 2024, has gone the closest to the Sun of any artificial object, with the goal of exploring the intense environment of the solar atmosphere. The Sun is the only object whose atmosphere we'll be able to go visiting for the foreseeable future, so sending a carefully shielded spacecraft into this dangerous area (it's hot, and electronics don't traditionally enjoy being exposed to plasma or charged particles) lets us figure out the details of how our own star launches material into the solar system around it. It is planned to continue operations ever closer to the solar surface until December 2025, but it has already revealed that the boundary between material bound to the Sun and material flung away from it is not totally spherical, but has outward protuberances in places, probably related to solar activity on the surface (fig. 2.2).

by its internal structure

If we can look deep into a star's heart, we can learn how it functions. In the innermost 20 percent (approximately) of the Sun's radius sits the core of the star (fig. 3.1). It's in this zone that the temperatures and densities are the highest, sitting at some 33 million Kelvin (60 million degrees Fahrenheit), ten times the density of lead, and holding about 50 percent of the mass of the Sun. It's in this profoundly inhospitable crucible that all the light that escapes the Sun is formed.

Once formed, the light in the core then has a long journey to escape the star. Bouncing into every single atom in its way, it is absorbed and re-radiated in random directions through what is known as the radiative zone. This zone—where no fusion can be found, but instead a chaotic scramble of light—extends outward to about 70 percent of the radius of the Sun. This is such a large volume to cover by random bouncing that it can take somewhere between 100,000 years and 1 million years for light to find a path out into the convection zone.

The trip through the convection zone isn't quite as arduous, as the material in this region is buoyantly lifted to the surface of the Sun through large cells. After the material has expanded and cooled in its rise toward the surface, the plasma sinks back into the depths of the Sun, completing a convective loop.

The "surface," or photosphere, of the star is marked as the distance from the core where light is finally able to freely stream away from it. Slight drops in temperature in the photosphere, known as sunspots, appear as dark patches against the surrounding brilliance. These appear due to the tangled paths of magnetic field lines bursting through the surface.

Fundamentally, all stars are operating in a delicate balance. Gravity would love to take the huge amount of material held in the star and crush it into a smaller space. But we see stars as stable objects in the skies, so some outward pressure must be successfully balancing that inward force.

Stars are made of plasma, and in many respects plasmas operate in the same way as gasses. Like most gasses, if you compress a plasma it will heat up. Gravity started this process of heating the

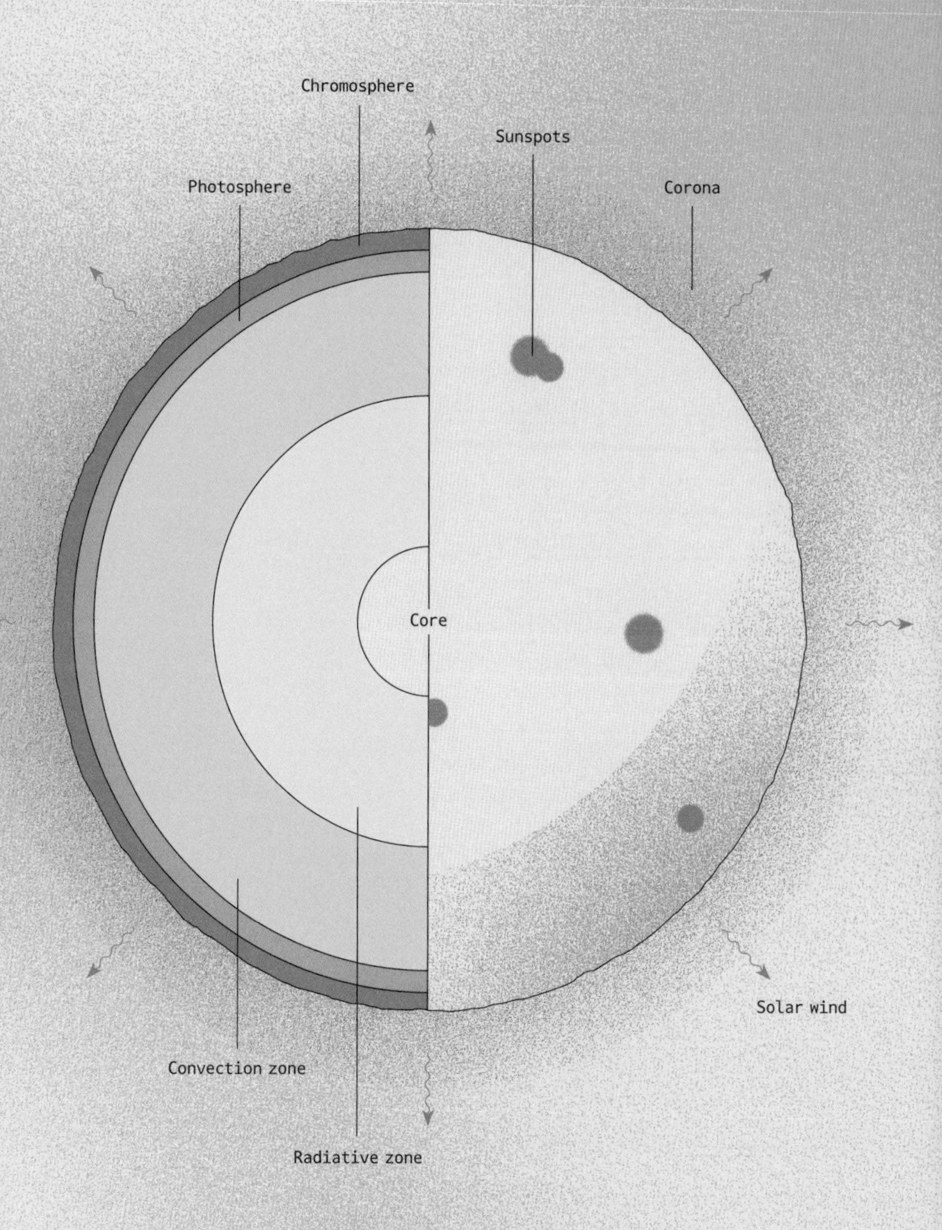

FIG. 3.1	Stars like our Sun have three major, distinct internal layers. All light is produced in the core, and takes a lengthy journey to reach the photosphere, where light can finally depart the star.

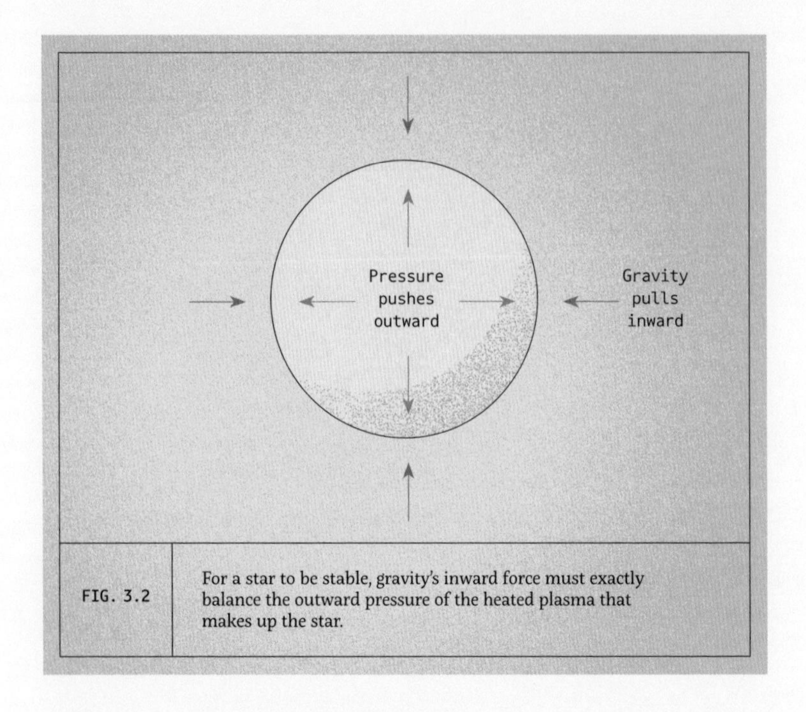

FIG. 3.2 For a star to be stable, gravity's inward force must exactly balance the outward pressure of the heated plasma that makes up the star.

star, but in order to effectively balance gravity we need also to be generating heat within the star (fig. 3.2).

In the core of a star, the density and temperatures are sufficiently high to provide a solution to our balancing problem: the fusion of hydrogen atoms into helium. Fusion is an interesting process—temperatures need to be outstandingly high, and atoms need to be so close together that they have a chance of slamming directly into each other at high speeds. Since the pathway from hydrogen to helium is a multistep process, there needs to be enough material in the vicinity to serve as the reservoir of ingredients required to keep the process going.

The steps of this process all involve slamming another proton—a hydrogen nucleus—into our attempt to build a helium atom. So, to complete this fusion process at all, we need a lot of hydrogen nuclei. Fortunately, most stars are made up overwhelmingly of hydrogen (fig. 3.3), with the majority of the rest comprising helium (a discovery we owe to Cecilia Payne-Gaposchkin's PhD thesis in 1925).

The end result is a singular atom of helium: two protons and two neutrons, built out of six protons, two of which are released back into the stellar core to participate in some other fusion reaction. Critically, the amount of mass held in one helium atom is not quite the same as the mass of four protons, and so a tiny amount (0.7 percent) of mass is converted into energy. Einstein's famous equation $E = mc^2$ tells us that we don't need a lot of mass to create a lot of energy—and so this process, while it has some strict requirements to get started, is quite efficient at generating energy, usually transported as light.

This constant generation of energy in the core of the star allows the plasma in the Sun to remain at such a high temperature that the mass of the entire star is not enough to compress the star. Gravity finds itself counterbalanced by the gas pressure generated from the heat created by fusion in the core. This stalemate is formally known as hydrostatic equilibrium, and all stable stars are found in this configuration.

While a single fusion reaction doesn't generate all that much heat, once we account for all the simultaneous reactions taking place throughout the core, there is enough to heat the plasma of the star, resist gravity, and glow into the vast dark of interstellar space.

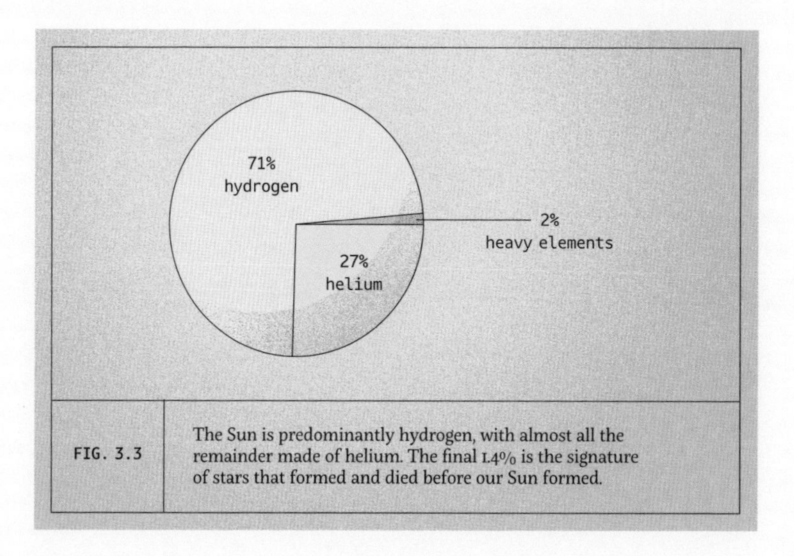

FIG. 3.3 The Sun is predominantly hydrogen, with almost all the remainder made of helium. The final 1.4% is the signature of stars that formed and died before our Sun formed.

during a solar eclipse

There is a particularly dramatic event during which we can get to know different aspects of our closest star, the Sun: a solar eclipse. It should be remembered that no part of a partial solar eclipse is safe to observe with the unprotected eye. However, if you're lucky enough to find yourself in the path of totality during a total solar eclipse—where the entire disk of the Sun is blocked out by the Moon for a few minutes—it is possible to view that portion of the eclipse without safety equipment, but taking great care (fig. 4.1).

During totality, two features of the Sun are visible to the naked eye which are not normally visible—most dramatically, the stellar corona. The corona appears as wide, white streamers of material flowing away from the star. It's usually invisible because it's much fainter than the surface of the Sun. It is, paradoxically, much hotter than the solar photosphere, and its faintness comes from how thinly spread the material within the corona is—it's much more diffuse than the solar surface. The temperature of the photosphere is a mere 5,772 Kelvin (9,930°F), compared to the casual 1–2 million Kelvin (1.8–3.6 million degrees Fahrenheit) of the corona. How the corona manages to be so profoundly heated relative to the surface of the Sun is still something of a scientific mystery, and is one of the main science goals of the Parker Solar Probe.

The other structure abruptly visible during a solar eclipse is the chromosphere, which glows a shocking pink color. This is the glow of ionized hydrogen. Like the corona, the chromosphere is usually too faint to be seen when the disk of the Sun is visible. It's a few thousand feet in height, and usually what is seen during a solar eclipse is not the main chromosphere but projections out from it, which are typically solar prominences.

There are three main classifications of solar eclipse, which can occur in any solar system. At the most fundamental, a solar eclipse occurs when an object—usually a moon—passes between an observer and the sun, blocking the light from the star.

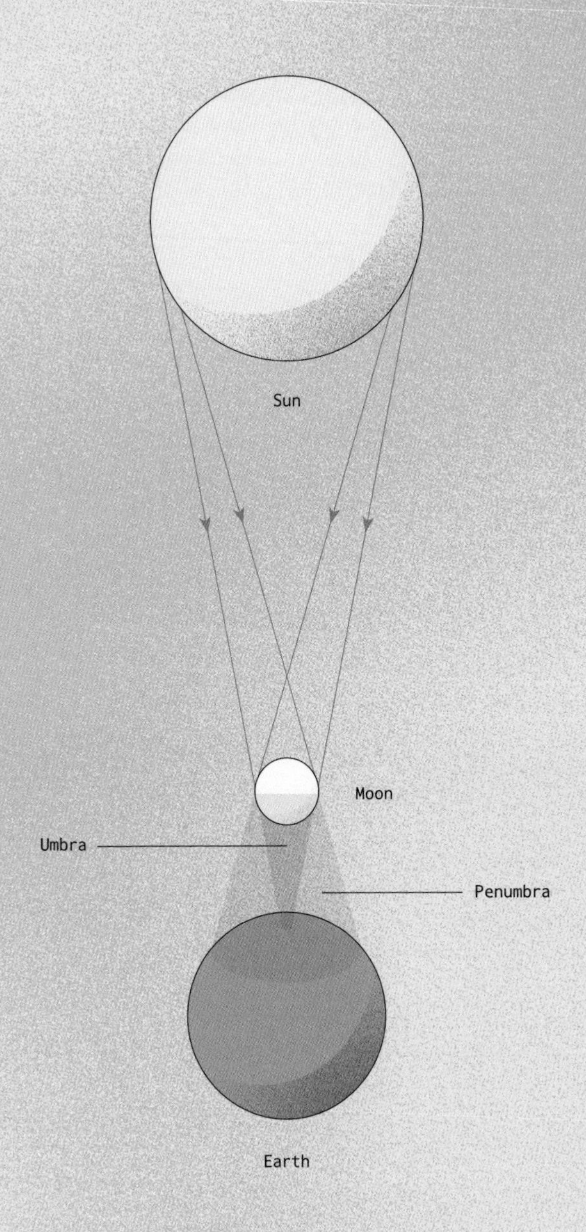

FIG. 4.1	During a total solar eclipse, it is safe to view the Sun directly; only the corona and the edge of the chromosphere are visible, with the bright disk of the Sun blocked by the Moon.

Total solar eclipses are quite rare. You need to have an object that is almost exactly the same size in the sky as the Sun—if they aren't the same size, it must be larger—and that object needs to pass precisely in front of the disk of the Sun. In our case on Earth, the Moon fulfills these requirements and can, in some months, pass directly in front of the Sun. In other months, because the Moon's orbit is not completely lined up with the Sun's path across the sky, the Moon misses the Sun entirely.

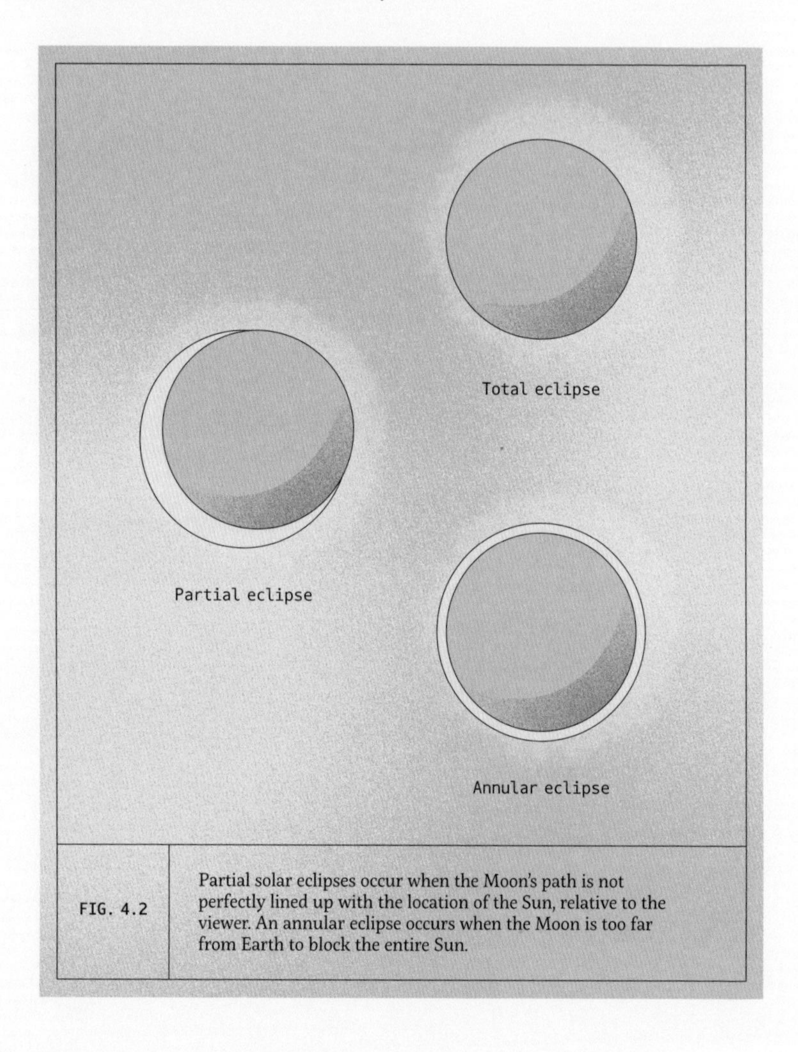

FIG. 4.2 Partial solar eclipses occur when the Moon's path is not perfectly lined up with the location of the Sun, relative to the viewer. An annular eclipse occurs when the Moon is too far from Earth to block the entire Sun.

Partial solar eclipses are the result when the moon or other object isn't lined up properly in the sky to fully block the Sun, and so a crescent Sun remains. Here the alignment is close, but not perfect. During a partial solar eclipse on Earth, pinhole cameras, colanders, and the gaps between tree leaves will cast crescent shadows, projecting an image of the eclipsed Sun. These are more common than total solar eclipses.

Finally, we have an annular eclipse. This is what happens when the object in question is not big enough to fully block the light from the Sun. What remains here is a ring of light surrounding the moon. We can see annular eclipses here on Earth: the Moon's orbit is oval-shaped, and if the Moon is unusually far away from Earth, it will appear smaller in the sky, and will not be large enough to fully block the Sun's photosphere. These kinds of eclipses are also unsafe to observe without protective gear. This is the only kind of eclipse visible from Mars. Both Phobos and Deimos, the moons of Mars, while quite close to the planet, are far too small to fully block the Sun, as seen from the Martian surface. Our robotic observers have captured images of Martian eclipses for us, but they are—at best—annular eclipses (fig. 4.2).

It seems to be quite rare to have a moon large enough to be able to block out the disk of the Sun so perfectly. Our Moon is an atypically large moon for the size of the planet we inhabit, so we're fortunate to be able to reap the esthetic benefits of a cosmic accident that formed the Moon billions of years ago. In the far future (millions of years away) Earth will also lose this cosmic vista. The Moon is receding from Earth at the rate of a few inches every year, and as this drift continues, the Moon, even at the closest point in its orbit, will only ever provide an annular eclipse.

PLATE 1

The last sliver of sunlight

In the moments just before and after totality, the faintest sliver of the Sun's disk glows brightly along one side, in what is called the "diamond ring" effect. This image was taken from NASA Glenn Research Center outside Cleveland, during the total solar eclipse on April 8, 2024.

by its mass

Learning about how much material is inside a star is a helpful step toward knowing how that star remains stable. All stars are a balance between the force of gravity inward and pressure from fusion resisting it, but what is interesting is that the balance can be different for every star. If a given star has a little more mass, then there's a greater force of gravity pressing inward. In turn, this increase in gravitational force means that if the star is going to remain stable, the outward pressure from the energy generated in its core must also increase.

Fortunately, there's a natural mechanism to achieve this. As gravity presses the star inward, it increases both the density at the core of the star and the temperature of the plasma. These are requirements that need to be met if the core of the star is going to accelerate how rapidly it is fusing hydrogen into helium. A balance can be restored in a more massive star by having it churn through the available hydrogen in its core at a faster rate than a lower mass star. This creates more energy in the core, and more resistive pressure, which settles the star back into a stable equilibrium (fig. 5.1).

The only downside to this solution, for the star, is that it's converting the available hydrogen in its core into helium much faster than a lower mass star is. Since that hydrogen fuels the fusion reaction, a higher mass star will deplete itself of hydrogen much sooner than its lower mass counterpart. Without the available fusion, the star will begin a complex end-of-life process much sooner than a lower mass star.

These statements work for any two stars of different masses. A star twice as massive as our own Sun will consume its fuel faster than the Sun, but by the same token, the Sun consumes its own fuel faster than a star with three-fourths its mass.

Very massive stars tend to be very bright stars. The extra energy being generated at the core means that the star produces more light. On top of the pure increase in light, more massive stars also tend to be physically larger. With a larger surface area, and a lot of light being produced in its core, a massive star will glow brightly, as long as the hydrogen holds out in the core. For a very massive star—say, sixty times the mass of the Sun—this fuel supply can last something like 3.5 million years. For the Sun, it's nearly 10 billion years.

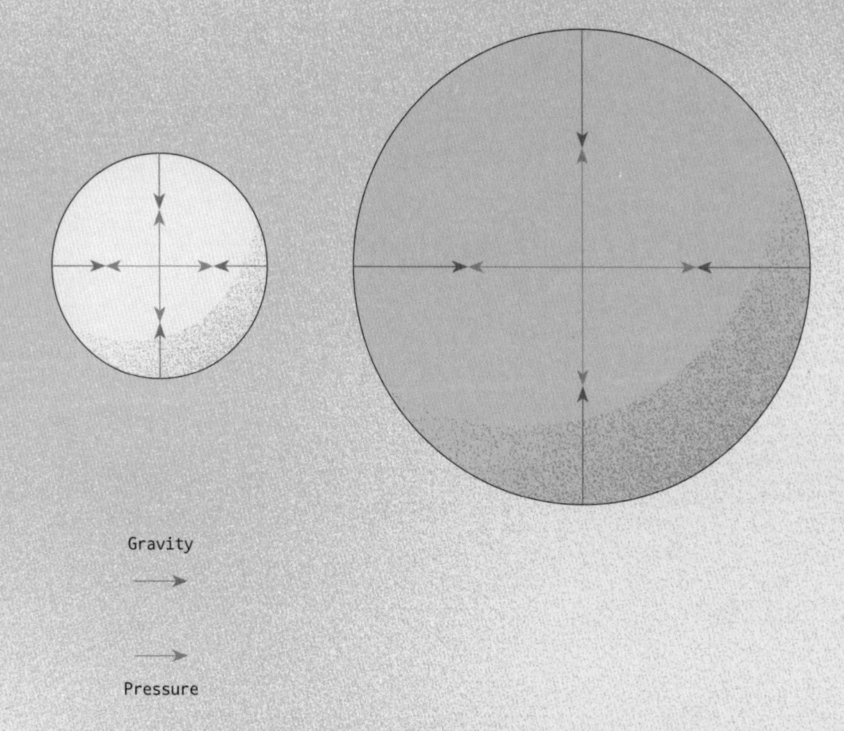

FIG.5.1	The more massive a star, the more gravity pulls inward. This is balanced by a higher pressure from rapid fusion. Low mass stars fuse less rapidly, balancing a weaker gravitational force.

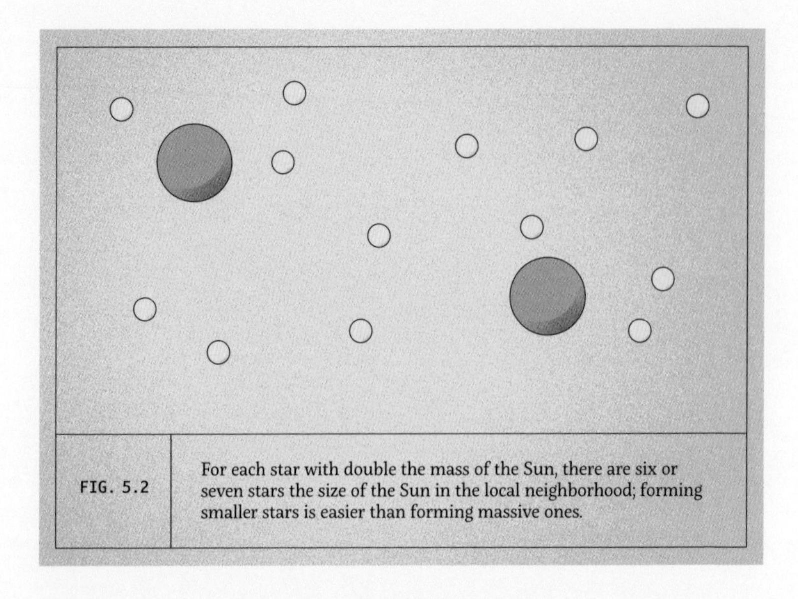

FIG. 5.2 For each star with double the mass of the Sun, there are six or seven stars the size of the Sun in the local neighborhood; forming smaller stars is easier than forming massive ones.

Mass underlies so many properties of a star that it's one of the fundamental pieces of information we need for every star we observe. With the mass, we can understand how long we should expect it to live, because we know something about how fast it runs through its hydrogen. It also tells us how bright we should expect it to be. Since we can measure one star's mass extremely well, we've used that star—our Sun—as a mass benchmark. Our own star weighs in at a very impressive 4.4×10^{30} lb (2.0×10^{30} kg) also known as one stellar mass. All other stars are measured in comparison, so a two stellar mass star has twice the mass of the Sun, at 8.8×10^{30} lb (4.0×10^{30} kg). This turned out to be not so unreasonable a metric, as the range of stellar masses in the present-day Universe lies somewhere between one-twelfth and 120 times the mass of the Sun (fig. 5.2).

To form a massive star is more challenging than to form a low mass star. A low mass star doesn't need so large a gas cloud to collapse down into a single object, and this is relatively easy to do. A high mass star, by contrast, needs to start with a much bigger reservoir of gas to fall inward on itself—and, critically, without splitting up into a series of smaller stars. The end result is that for every very high mass star, we tend to have a LOT more low mass stars. In the neighborhood around the Sun, stars like the Sun are about 6.5 times more common than stars with 1.5–2 times the Sun's mass.

The low mass end of things is also interesting. While the high mass end has limits based on how tricky it is to form such a star in the first place, the low mass end has a different limit. The lowest mass stars are only barely hot enough and dense enough in their cores to be able to support any fusion at all. About one-twelfth the mass of the Sun is the lowest mass an object can be and still fuse hydrogen into helium (fig. 5.3). Following the inverse relation of the high mass stars, we can infer that they turn their hydrogen into helium at a very slow pace—and indeed it seems that these stars are likely to live for trillions of years without running out of fuel.

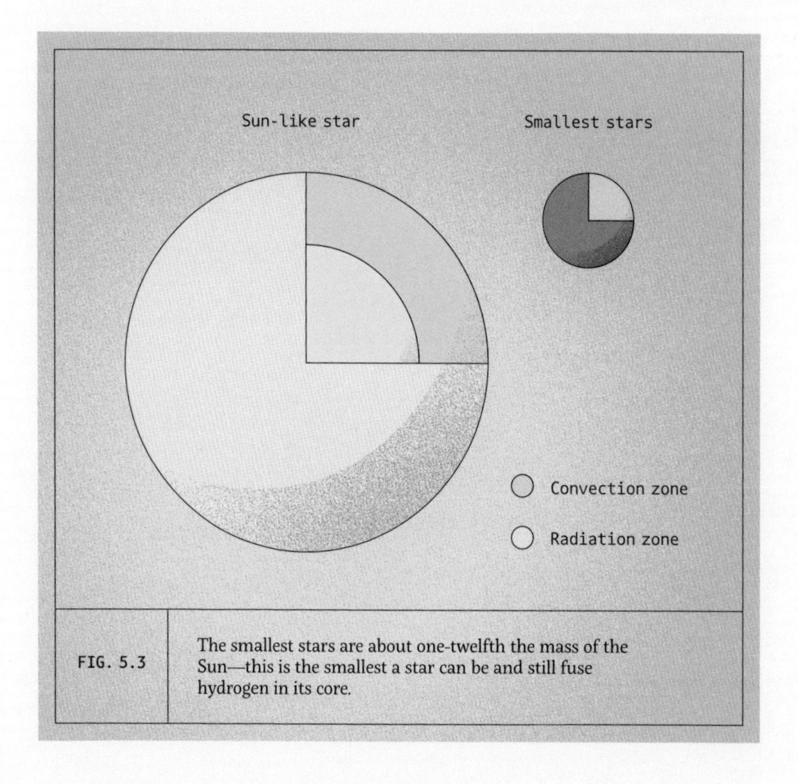

FIG. 5.3 The smallest stars are about one-twelfth the mass of the Sun—this is the smallest a star can be and still fuse hydrogen in its core.

[SIX] Know a star...

by its color

Careful inspection of a star's color can teach us a lot. Looking at the night sky, a couple of stars are more than bright points of white light: they're red. Betelgeuse, in the shoulder of the constellation of Orion, and Antares, in Scorpio, are both distinctly redder than the average star. If we look more carefully, and with technology more suited to the task, we can see that all stars have a color associated with them—and it tells us directly about the temperature of the surface of the star.

Stars are classified as blackbody radiators, which is the technical term for something that's glowing because it's heated to a high temperature. There aren't a lot of blackbody radiators in modern daily life—which is probably for the best, because having objects in your house so hot they glow is a great way to increase your risk of severe burns. If you have an old coil electric stovetop, or an older incandescent light bulb, those would meet the criteria. Molten metal also glows this way. Heated to a yellow-white hue, we can see that the material is hot well before we get close enough to feel the heat coming off of it. The hotter the material, the more white it looks—there's science behind "red hot" being not quite as intense as "white hot."

If you examine something that's glowing under the power of its own internal temperature, and look at what colors of light it produces, you can create a graph that shows the spectrum of light, and how much of each color is produced. For objects glowing due to their own heat, a very specific curve is created—the blackbody curve. It's a wide curve, with one wavelength of light more prominent than the others (fig. 6.1).

For the Sun, this peak wavelength sits at a furiously lime green color. Our eyes don't perceive this as the color of sunlight because our eyes average out all the frequently produced colors. If you average across the range of visible light, we interpret this as "white" light, even though the Sun is really producing all sorts of colors. We do get to see these colors individually when a rainbow appears, and the sunlight is broken out into all its constituent colors.

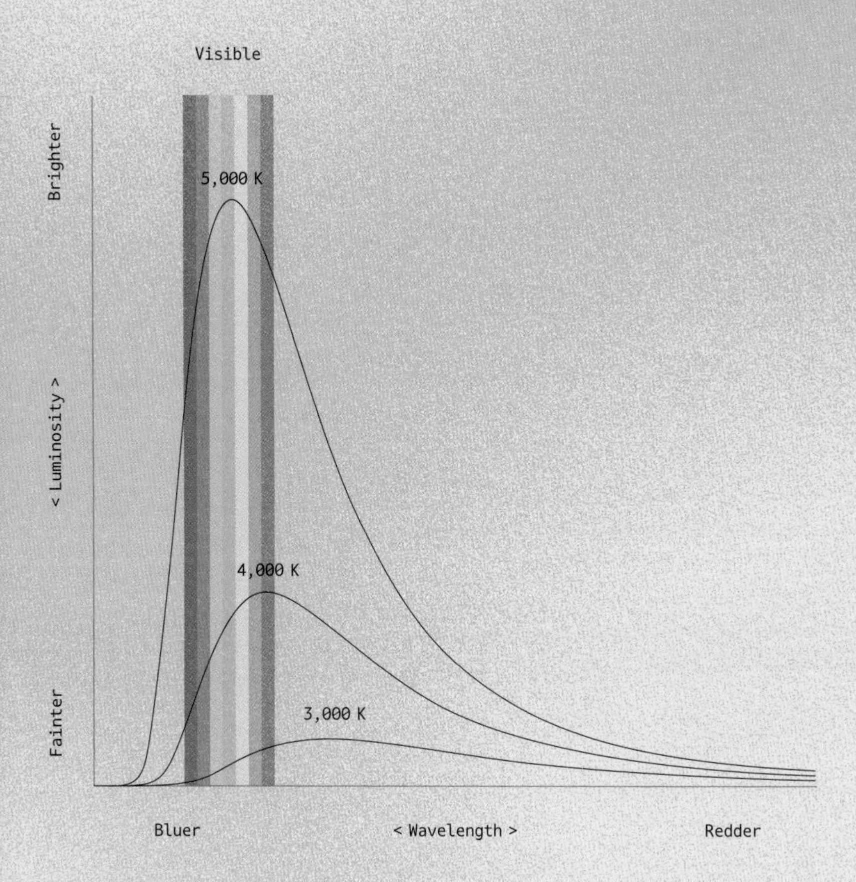

FIG. 6.1	Blackbody curves for stars of different temperatures. The further to the blue end the most common color produced by a star is, the hotter the surface temperature of the star must be.

This peak wavelength is critical to understanding stars, because the bluer the wavelength, the hotter the object is. Redder peak wavelengths mean that the surface is cooler—and we see them as redder, even averaged out, because they simply do not produce much blue light. Antares and Betelgeuse, therefore, must have relatively cool surfaces (for a star), to show up as red as they do in our skies. By contrast, Sirius, the brightest star in the night sky, is a brilliant white, and is indeed both bluer and hotter than either Arcturus

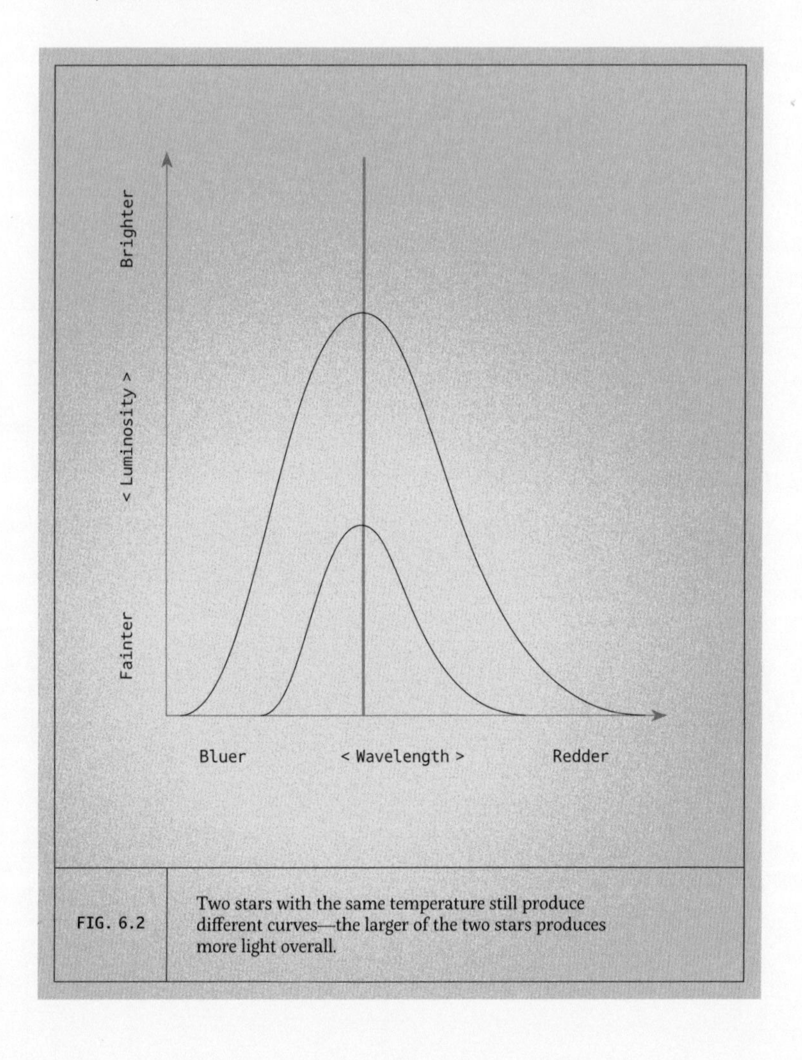

FIG. 6.2 Two stars with the same temperature still produce different curves—the larger of the two stars produces more light overall.

or Betelgeuse. All stars produce light this way, and so if you can split the light from a star apart, and figure out what color of light it produces the most of, you can learn what the temperature at the surface of that star is.

This bluer color with hotter stars also matches up well with our understanding of the interiors of higher mass stars. Since these are fusing hydrogen at a much faster pace to compensate for gravity compressing their cores inward, they're producing a lot more energy, and by the time that energy reaches the surface, the star is much hotter overall, and therefore bluer.

Color only tells us about surface temperature, though. It doesn't actually tell us anything about the size of the star. There is generally a correlation between higher mass and higher temperatures, but there are exceptions here—sometimes you have a relatively massive star with a cool surface. This is the first hint that stars do not always stay in equilibrium the way we've described so far. On this diagram we can actually distinguish between a star's low surface temperature because of its low mass, and a low surface temperature because the star has expanded to tremendous sizes—not very warm, but very luminous. If the star has a lot of mass, and it's got a red surface, it will be physically much larger than a star that doesn't have a lot of mass. And if it's much larger, then its surface area is much bigger, which means that overall, it's producing more light, even if it's producing the same wavelengths as a smaller star (fig. 6.2).

Betelgeuse is a good example again. It's very visibly red, but part of why it's so bright is because it's physically enormous. It's about one thousand times the size of our own Sun. That's not a size that a star with half the mass of the Sun can achieve, so it's got a low surface temperature for other reasons—we'll come to those in later sections.

PLATE 2

A multicolored gravitational collection

The open cluster NGC 299 shows
a range of stellar colors, all formed out
of a single cloud of dust and gas. Located
within the Small Magellanic Cloud,
around 200,000 light-years away, clusters
such as this one offer astronomers
a cosmic laboratory in which to study
the evolution of stars.

by its wobble

While some stars form in relative isolation, like our Sun, it's fairly common for stars to form in pairs—some 45 percent of Sun-like stars seem to have companions, and that fraction goes up as the mass of the star increases. To find these companions, we can watch for unexpected wiggling in their light.

We usually think about smaller objects orbiting heavier ones, as the Moon orbits the Earth and the Earth orbits the Sun. For systems like these where one object is much more massive than the other, this is approximately true. Technically, though, any two objects which are gravitationally tied to each other orbit around their common center of mass. For the Earth and the Sun, the center of mass between the two is approximately still in the center of the Sun, but not exactly the center of the Sun. But if you have two equal mass stars orbiting their center of mass, they'll both orbit a point in the unoccupied space exactly halfway between those two stars.

If the stars are near enough to us, and our perspective from Earth allows us a bird's-eye view of the looping path both stars take, we can, with a precise enough telescope and a long enough time to watch, see the stars move around each other. With this information, we can trace out how the stars move—exactly what their orbit is. The most critical piece of data we can get here is how long it takes the two stars to trace out one loop of their own orbits. This timeline will let us determine the mass of the two stars, which, in turn, will let us figure out their lifetime, among many other things.

There's a relationship between the length of time it takes any object to complete one orbit and the total amount of mass in the system. This comes directly from our understanding of the laws of gravity. We also need to know how far apart the two stars are, but if we can see them moving, and estimate how far away they are from us, we ought to be able to get all the pieces we need (fig. 7.1).

Most of the time, geometry isn't in our favor—there are a lot more ways for things to line up without getting a bird's-eye view than there are ways for us to get that perfect angle. In those

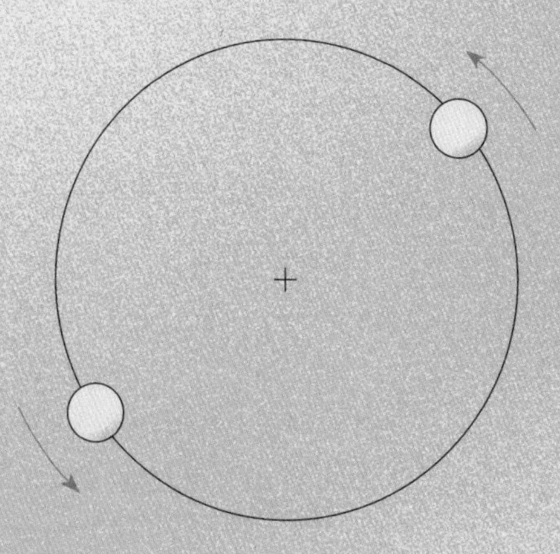

Equal mass stars

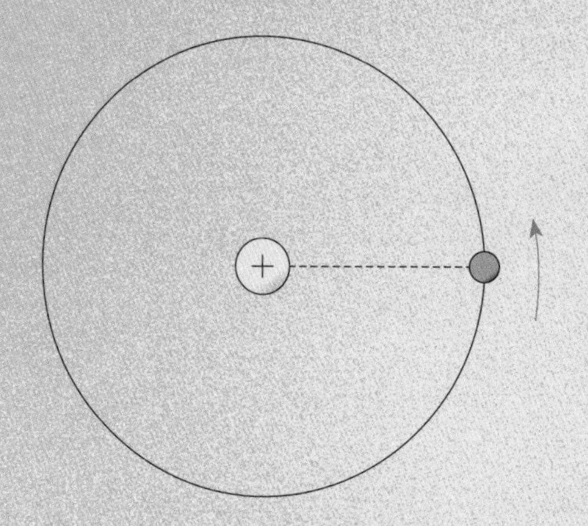

Earth and Sun

FIG. 7.1	When objects have unequal masses, they approximately orbit the more massive object, but technically it's the center of mass between them. When the objects are equal in mass, they orbit a point in the empty space exactly halfway between them.

cases, we can use the fact that the orbit of the stars means that each star spends half its time coming toward us, and half its time traveling away from us.

Much like the Doppler shift of a siren or car horn changing from a high to a low tone as it passes us, light does the same thing, but here it's the distance between the waves of light rather than sound. When the waves of light are closer together, it's bluer light that reaches us, and when they are further apart, it's redder light. Uninventively, we've called this "blueshift" and "redshift" and it can happen with any motion of a glowing object relative to our vantage point on Earth (fig. 7.2).

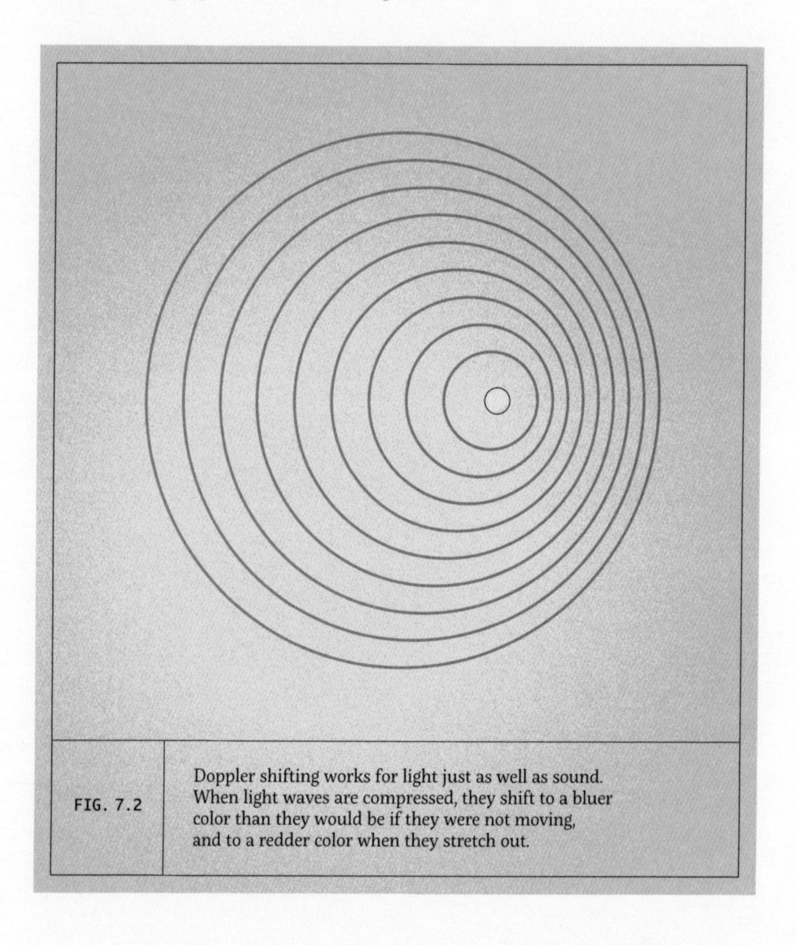

FIG. 7.2 Doppler shifting works for light just as well as sound. When light waves are compressed, they shift to a bluer color than they would be if they were not moving, and to a redder color when they stretch out.

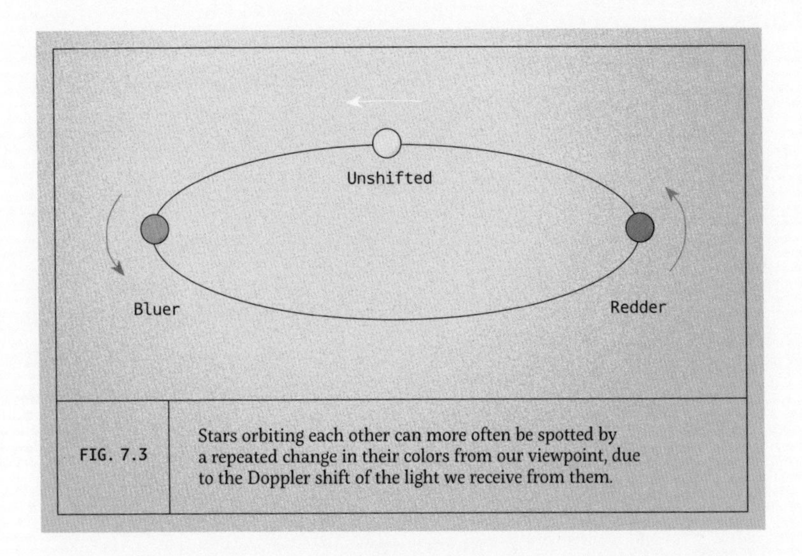

FIG. 7.3 Stars orbiting each other can more often be spotted by a repeated change in their colors from our viewpoint, due to the Doppler shift of the light we receive from them.

Stars orbiting each other cycle through a blueshift, then a redshift, then a blueshift, then a redshift again. The length of time it takes for the shifts to switch back and forth gives us the same information about how long the stars take to orbit each other that we would get if we could simply watch them go around. And instead of being able to see their separations in the sky to estimate how far apart they are, we can look at how fast they're moving. With a travel time and a speed, we can figure out how much distance the orbit must take them along, and that in turn lets us figure out how far from the center of their orbit they must be. All without being able to see the stars move (fig. 7.3)!

The downside to this approach is that we usually can't calculate the individual masses of the stars—we get the sum of the two. So if we learn from this approach that the total mass of the two stars is 3 solar masses, that could mean that one of them is 1 solar mass and the other is 2, or that both of them are 1.5 solar masses, or it could mean that one is 1.4 and the other is 1.6 solar masses. Without extra information about the two stars, we can't get the exact masses of them, but given that we couldn't see either star clearly on its own, we're doing quite well to get the sum of both.

by magnitude

We can also measure the brilliance of the stars, continuing work begun in antiquity. Ancient astronomers observed the skies with the most advanced telescopes they had available to them: their eyes. So when they were beginning to create catalogs of the locations of the stars in the sky, they began to note not just where the stars were, but their brightness. Helped by a profound lack of light pollution prior to the invention of industrial lighting, these early record-keepers had a lot of stars to make a note of. The first known record of the stars is a catalog created by the Greek astronomer Hipparchus, probably somewhere around 135 BCE, and for many years the only hint of its existence was in references to it elsewhere. In 2022, however, a transcription of a portion of it was found hidden on a parchment which had been scraped clean and reused sometime between 800 and 900 CE.

While the ancient Greeks did have mechanical tools to help them measure locations in the sky, the best metric these astronomers had for measuring brightness were their eyes. How bright does the star appear to the eye? Hipparchus took an initial note of these brightnesses, but it was in Ptolemy's *Almagest*, published in about 137 CE, that stars were classified into a set of six brightness categories. The "brightest"—first magnitude stars—were followed by fainter second magnitude stars, down to sixth magnitude stars, which is the limit of what the unaided eye can see in dark skies.

With technological advancements came the need to standardize things slightly. In 1850, instead of putting stars simply within one of six classifications, it was recognized that a first magnitude star was about 2.5 times brighter than a second magnitude star. In fact, the whole magnitude system developed by the ancient Greeks turned out to be a logarithmic one—five magnitudes meant a change in brightness of a factor of 100 (fig. 8.1).

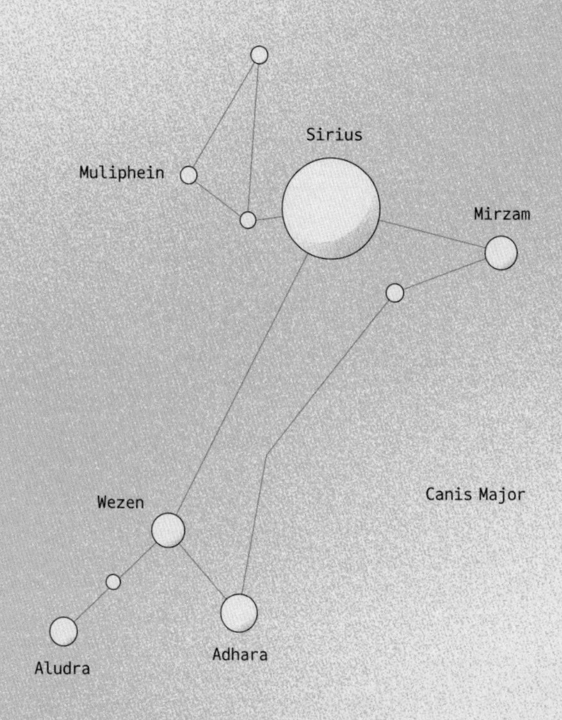

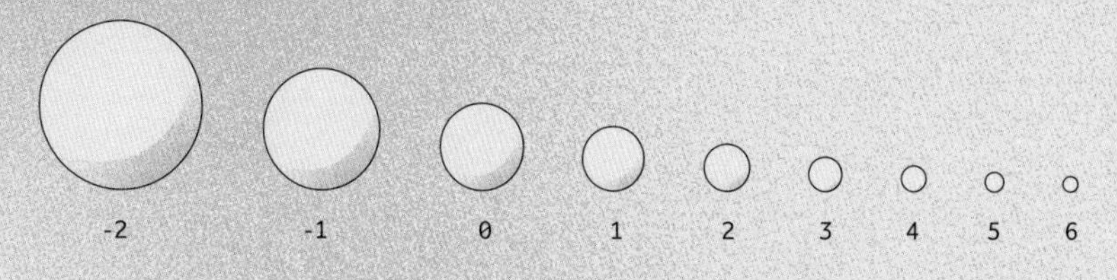

FIG. 8.1	The ancient Greeks created a system of classifying the apparent brightness of the stars in the night sky, from the faintest stars observable (sixth magnitude) up to the brightest (first magnitude). Here, Canus Major is illustrated with the stars to scale by their magnitudes.

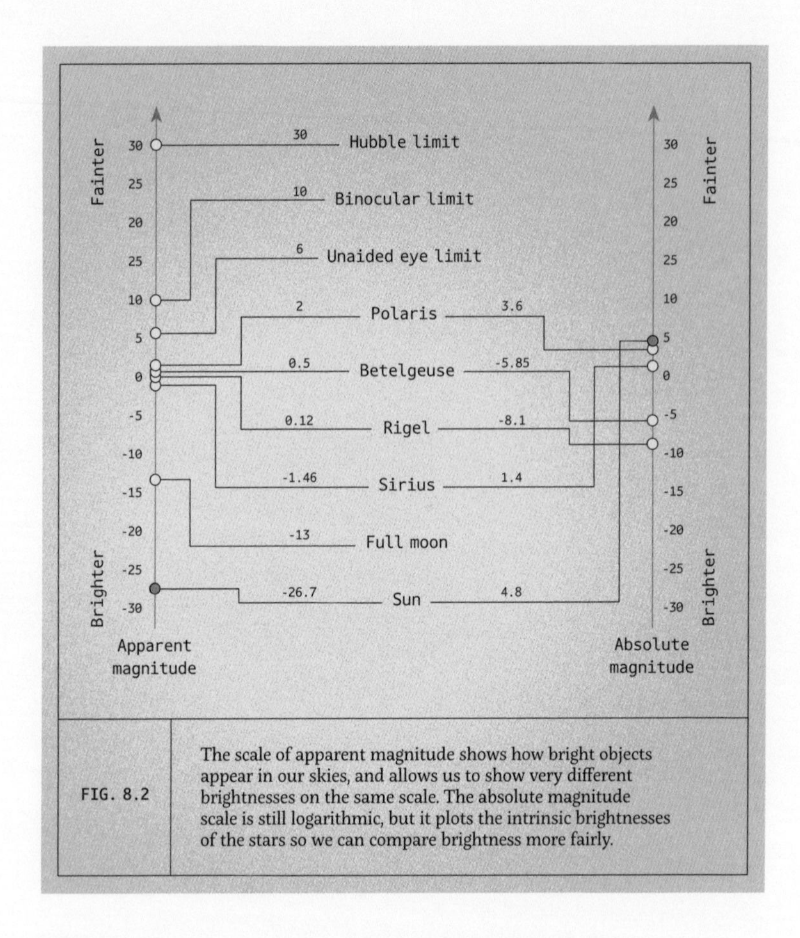

FIG. 8.2 The scale of apparent magnitude shows how bright objects appear in our skies, and allows us to show very different brightnesses on the same scale. The absolute magnitude scale is still logarithmic, but it plots the intrinsic brightnesses of the stars so we can compare brightness more fairly.

The fact that the magnitude system was logarithmic meant that you could place things of extremely different brightnesses on the same scale without the numbers getting unreasonable. And standardizing the scale meant that we could break out of the round number magnitudes, and could start classifying stars as partial magnitudes: 3.5 is fainter than 3.2, both of which are fainter than a magnitude 1.2 star.

Unfortunately, when we standardized the brightnesses, some of our formerly "first magnitude" objects weren't all the same brightness, and some of them needed to be placed into a brighter-than-first magnitude category. The only direction to go was negative.

So, much to the chagrin of every single astronomer, the brightest objects in the night sky have negative magnitudes. As seen from Earth, the Sun has a magnitude of -26.7 (fig. 8.2).

This is an apparent magnitude—how bright it appears in the sky to us, from our vantage point. It's influenced by two things: first, how intrinsically bright the object is, and second, how close it is to us. The Sun is not intrinsically extremely bright on an astrophysical scale, so its apparent magnitude of -26.7 is cheating by virtue of being incredibly close to us.

In further standardizations, we've also developed a system called the "absolute magnitude," which is the magnitude we would observe if we could put every object at the same distance from Earth, and observe them all on an equal footing. We've somewhat arbitrarily chosen 32.6 light years (10 parsecs) as the distance to use as our baseline, but in principle any distance could work as long as we were consistent about it. Absolute magnitude thus is a way of correcting for distance, and allowing us to compare the intrinsic brightnesses of objects to each other. The absolute magnitude of the Sun is +4.83. This means that by virtue of its nearness, it appears about 4 trillion (4.1×10^{12}) times brighter than it would if it were 32.6 light years away from us.

Sirius, the brightest star in our night sky, is pulling a similar, though not so extreme trick. It has an apparent magnitude of -1.46, but because it's also relatively close to us, the absolute magnitude of the star is +1.4, fourteen times fainter than it appears in our skies. By contrast, the star Rigel, in the constellation of Orion, has an apparent magnitude of +0.12—it's the eighth brightest star in the sky, including the Sun in our math. However, its absolute magnitude is -8.1. Because this star is further away than our 32.6 light year baseline, we're actually seeing it as fainter than it would be: some 1,900 times fainter, in fact.

All the stars we can observe have their brightnesses modified by the twin settings of intrinsic brightness and distance. Intrinsically faint stars can look brighter by virtue of being closer to us, and intrinsically bright stars can look fainter if they are distant from us. The two combine to give our unique view of the stars.

[NINE] Know a star...

by its birth

Understanding the formation of the star itself allows us to begin at the very beginning. All stars form in the same way. They begin their lives as a tremendous cloud of gas, bearing no resemblance at all to the blistering plasma of a fully-fledged star. But a cloud of gas with enough mass to make a star is the only ingredient that we seem to need in order for a star to form—though there are a few other criteria we'll need to fulfill. The first is that the gas has to be relatively cool. If the gas is too hot, then gravity can spend all the time it likes pulling inward on it, but the temperature of the gas will be high enough to resist that inward pressure. Secondly, even with a cooler gas cloud, we need time.

Either the cloud of gas will eventually fall inward on itself due to the constant compressive influence of gravity, or something can nudge it—some nearby explosion might reach our gas cloud and compress it just enough to cause instability in the cloud, which will lead to the start of a collapse. In either case, gravity is the key once collapse begins, and we're on the way to forming a brand new star.

In a simpler world, this gas cloud might be able to simply collapse inward on itself equally in all directions, but that's not the kind of straightforward Universe we live in. In actuality, all clouds of gas are in motion, and when the gas cloud begins to collapse in on itself, a property called angular momentum becomes quite important. The most familiar example of this is figure skaters, who can begin a spin with their arms extended, and then pull their arms inward and increase their rate of spin impressively. To apply this to stars, we just need to scale everything up (fig. 9.1).

For a cloud of gas beginning to collapse, it starts this process at the largest extent it will cover. As gravity pulls the cloud inward, its spin should increase from whatever faint drift the cloud had to start with. As this drift builds into solid rotation, the cloud of gas will develop a general rotation axis, around which all the gas is spinning. Along that axis, it is relatively easy for material to fall inward, simply following gravity "down" to the center of the cloud.

FIG. 9.1

A star, prior to the start of fusion in its core, is called a protostar, and it collapses down out of a large cloud of gas and dust into a protoplanetary disk, which takes about 50 million years to complete for a Sun-like star.

However, for material found 90 degrees perpendicular to the axis of rotation, while gravity is still pulling inward, the collapse of material into the center is gradually and increasingly resisted by the centrifugal force. Some of this material finds itself in a stable orbit, in a disk surrounding a dense central nugget, where the majority of the gas cloud has coalesced (fig. 9.1).

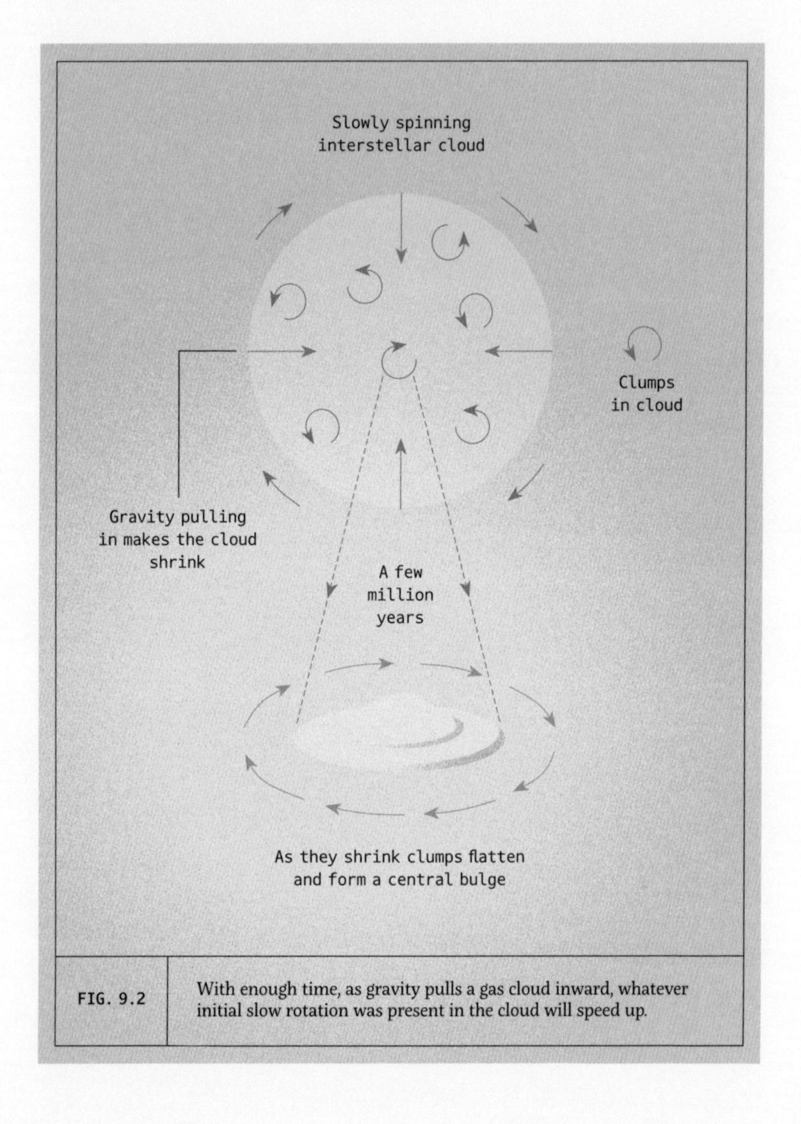

FIG. 9.2 With enough time, as gravity pulls a gas cloud inward, whatever initial slow rotation was present in the cloud will speed up.

At this point we've created a protostar. The object in the center of our former gas cloud is hot, having been compressed through gravity to a sufficient degree that the gas pressure has begun to heat the material, but it's not yet hot enough to begin fusion. The surrounding material, known as a protoplanetary disk, is where any future planets may form.

The heat from the protostar will now begin to destroy the protoplanetary disk, evaporating away any lightweight materials and beginning to generate a protostellar wind, which will also push heavier materials away from the emerging star. Once it clears the immediate area around itself of gas and dust, the protostar can no longer gain any further mass, and will slowly collapse—until its core heats up enough to trigger the first fusion reactions. At that point, the protostar transitions to being a fully fledged star, and will enter a period of stability for millions or billions of years, depending on how much mass it managed to build up in the protostar phase.

The whole collapsing process will take somewhere between 1 million and 100 million years to complete, depending on the mass of the star that's being formed—more massive stars collapse faster than low mass stars (fig. 9.2).

There's no rule that says one cloud must collapse down into a single star—and in fact it seems likely that most do not, given how frequently we see stars in clusters or binary pairs. An exceptionally large cloud of gas, which all begins to collapse at once, can easily form hundreds of stars. For instance, the Pleiades star cluster contains a few bright stars easily visible by eye, but more than 1,000 other stars are part of that one cluster.

PLATE 3

A star begins in dramatic fashion

This protostar is seen in infrared light by the James Webb Space Telescope. The narrow, horizontal dark band at the center of the image is the protoplanetary disk, with the cones of light vertically exposed to the light of the forming star.

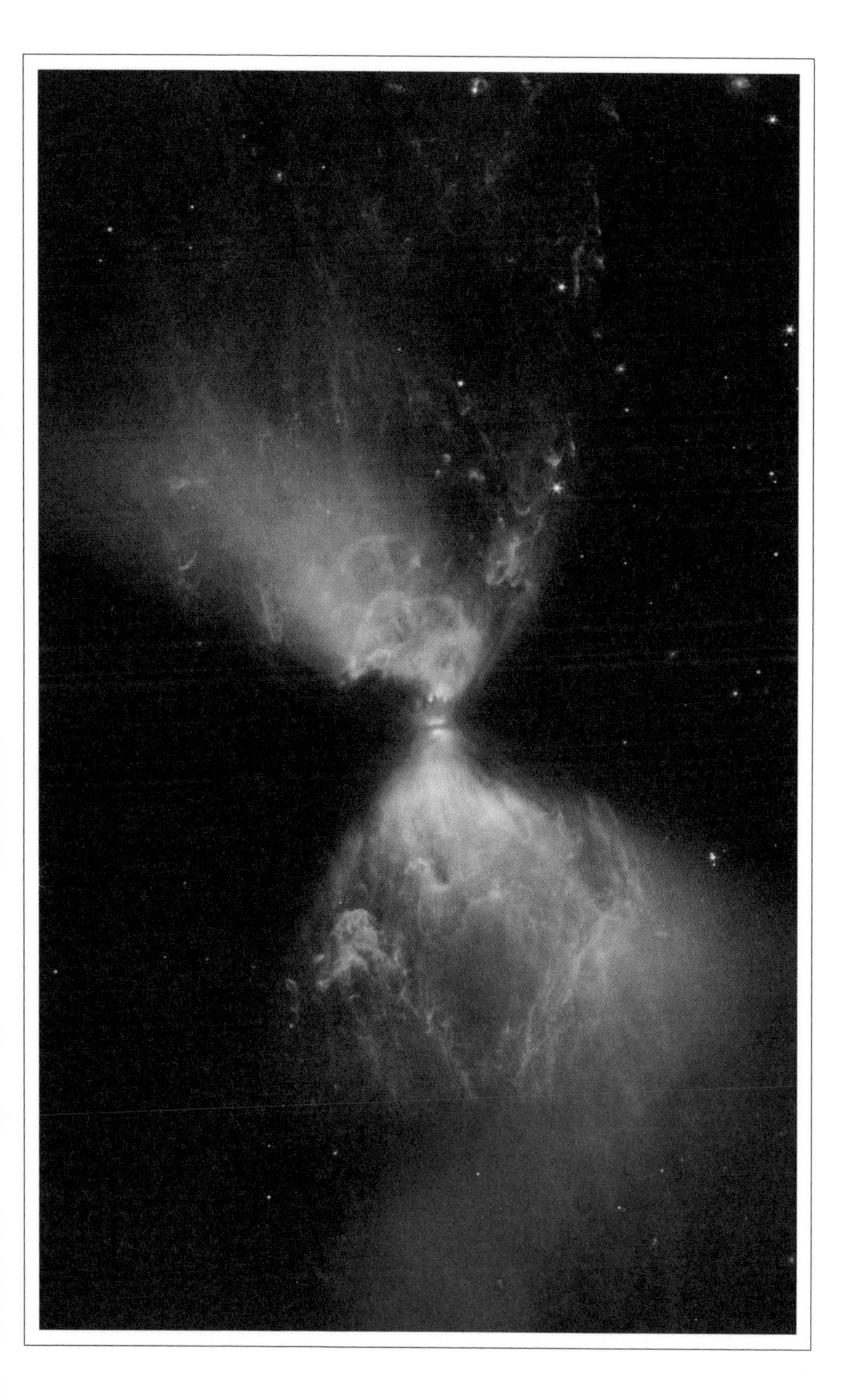

as the host of planets

The process of forming planets around a star takes us from swirling gas to discrete worlds. The protoplanetary disk is more than just the 0.1 percent of leftover material that didn't make it into a star. In many cases, while the protostar is working on becoming a star, the protoplanetary disk is busy transforming from a vague haze of gas and dust into something more elaborate.

Inside the protoplanetary disk, small chunks of material are beginning to form. Close to the protostar, where the heat from it is stronger, the material that condenses down into chunks is mostly made out of rocky substances, while further out small pieces of both rock and ice can condense. This can extend beyond water ice and into materials that would be gasses on Earth: methane, nitrogen, and ammonia are also in the mixture.

If we wait a few million years, this chaotic mess of small chunks of material will have built itself up into a slightly less chaotic pile of somewhat larger objects. These will become our asteroids and the seeds of what will eventually be planets, but for now they get the moniker of a "planetesimal." Planetesimals are considered to be objects less than 60 miles (100 kilometers) in diameter, but usually more than ½ mile (1 kilometer). At this threshold, the planetesimal is able to start wielding its own gravity to gather more materials to itself. If it's close to the star, it will simply have to run into other objects to grow, but if it's out in the distant realms of the solar system, it will be able to draw gas and ice inward through gravity. Once the planetesimal grows past around 60 miles (96 kilometers) in size, it's officially a protoplanet (fig. 10.1).

Meanwhile, the protostar will have formed, clearing out the inner solar system in a few tens of millions of years. This will end the formation of planetesimals and ensure that the only way for larger rocky bodies to grow is for the existing rocky planetesimals to slam heartily into each other. Fifty million years after the start of fusion in our Sun, most of the planets had probably formed. Jupiter likely formed relatively quickly and grew rapidly, which resulted in a planet that has more than double the mass of the rest of the planets combined.

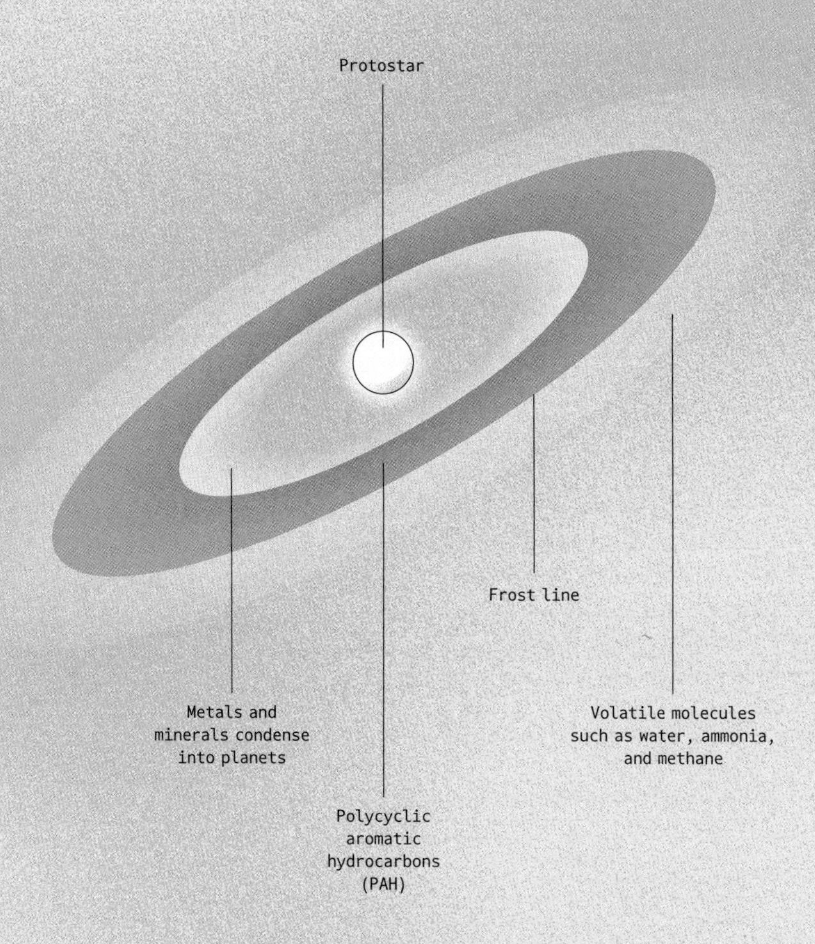

FIG. 10.1	Protoplanets form out of the protoplanetary disk, from the 0.1% of material that didn't make it into the star. They grow over tens of millions of years, colliding with other planetesimals to form the planets we know today.

In our solar system, it seems that the outer planets temporarily became unstable in their previous orbits, which had been closer to the Sun. They shifted out to their current orbits about 550 million years after fusion began in the Sun, which may have been responsible for a particularly unpleasant case of large asteroids hitting all the rocky planets. Overall, though, our solar system remained relatively orderly. All the rocky planets are in the inner solar system, and then we have a sequence of gas and ice giant planets further out.

You might think, with eight planets in our solar system, from the barren surface of Mercury to the equally barren but atmospherically oppressive Venus, to the winds of Neptune and the rings of Saturn, that we have covered the options for what planets could look like. But, having now investigated lots of stars and their planets over the last few decades, we've learned that many extrasolar systems look nothing like ours.

Many solar systems have planets we've dubbed "hot Jupiters." These are Jupiter-sized planets, but frequently orbiting closer to their star than Mercury orbits around our Sun. Blisteringly hot, and often orbiting in as little as four days, these are truly alien worlds, actively evaporating away their own atmospheres. They are wildly inhospitable, with atmospheric temperatures of thousands of degrees (fig. 10.2).

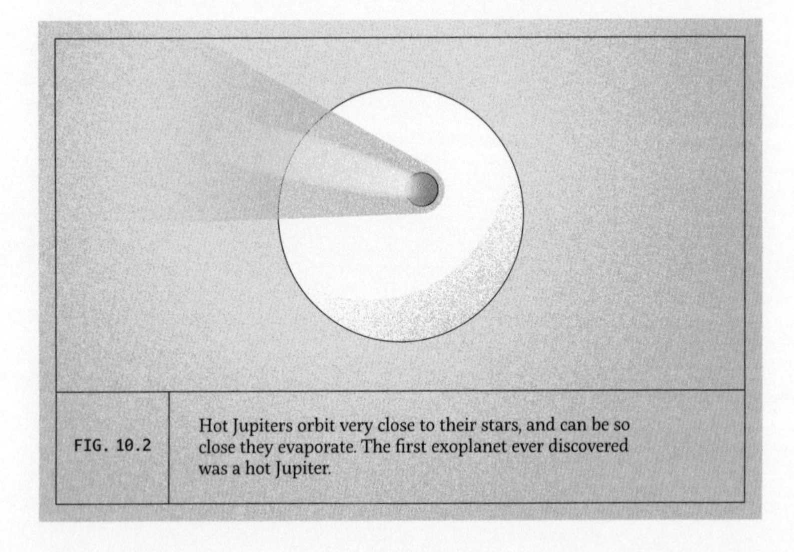

FIG. 10.2 Hot Jupiters orbit very close to their stars, and can be so close they evaporate. The first exoplanet ever discovered was a hot Jupiter.

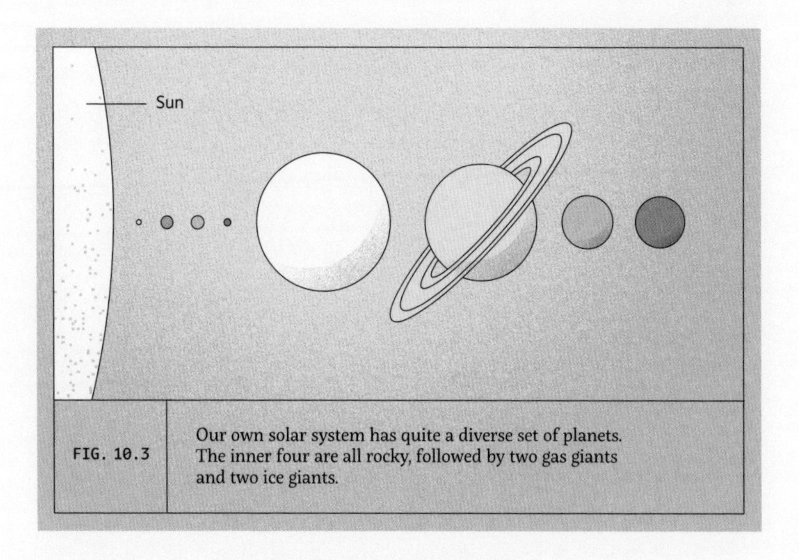

FIG. 10.3 Our own solar system has quite a diverse set of planets. The inner four are all rocky, followed by two gas giants and two ice giants.

There's another class of planet not seen in our solar system, which the planet-hunting telescopes Kepler and TESS have brought to light. These are planets that fill the gap between the Earth-like planet and the Neptune-like planet, dubbed super-Earths or mini-Neptunes depending on which we think they're more likely to resemble. Earth is the largest of the rocky worlds, and the smallest of the gas giant planets is Uranus, with about 14.5 times the Earth's mass, so there is a gap in planet sizes in our solar system. Other solar systems have shown us that it's not impossible to form planets in this gap; they just did not form in our solar system (fig. 10.3).

They're a puzzling class of planet, because planets between an Earth mass and the 14 Earth masses of Uranus also exist at the boundaries of how we understand planets to form. Some of them may be rocky, giant versions of Earth. Some may be huge Mercury analogs, devoid of any kind of atmosphere. Some may truly be tiny Neptunes: small, atmospherically laden worlds with no real surface to speak of. Some could be covered in water, with an oppressively heavy, thick atmosphere shrouding the watery surface. Some, like our current understanding of 55 Cancri e, may host a huge lava ocean. In all cases, the quest to understand these planets from afar is ongoing.

[ELEVEN] Know a star...

by the aurorae

There's an unusual glow in some planetary atmospheres: the aurorae. Seeing the aurorae in our atmosphere is proof that we live around a dynamic star. The aurorae themselves are the result of the solar wind—a constant stream of charged particles flowing outward from the Sun's surface through the entire solar system—encountering our atmosphere.

Usually, our atmosphere is fairly protected from the solar wind, because of the presence of a magnetic field surrounding the Earth. However, at the northern and southern magnetic poles, there are paths for the particles from the solar wind to find their way down to the atmosphere and interact with the gas it contains, causing it to glow. This is why the aurorae are more easily seen closer to the magnetic poles. And if the solar wind were completely constant, with the Sun a fully stable entity in our solar system, that might be the end of the story.

However, the solar wind isn't constant, and there are times when it becomes more dense. With any such "solar storm" comes a dramatic increase in the visibility of the aurorae.

The most dramatic solar storms occur as a result of a "coronal mass ejection" on the Sun. This is a large amount of material abruptly ejected from the surface of the star, and it takes about three days to arrive at Earth, if the ejection happens to be pointed our way. Most CMEs are not pointing in our direction, and simply make for dramatic imagery from any of our solar monitoring spacecraft. Since we have quite a number of these spacecraft, these days we generally have a three-day warning before a solar storm arrives (fig. 11.1).

Coronal mass ejections are most common when the magnetic field of the Sun is particularly tangled. This tangling happens on an eleven-year cadence, so if you're keen to go viewing them, there's a period of a few years every eleven years when they're most likely.

When the particles released in a CME reach our atmosphere, either by bending the magnetic field to expose the northern and southern atmosphere, or simply by tangling themselves in the magnetic field lines and spiraling inward, they will continue to

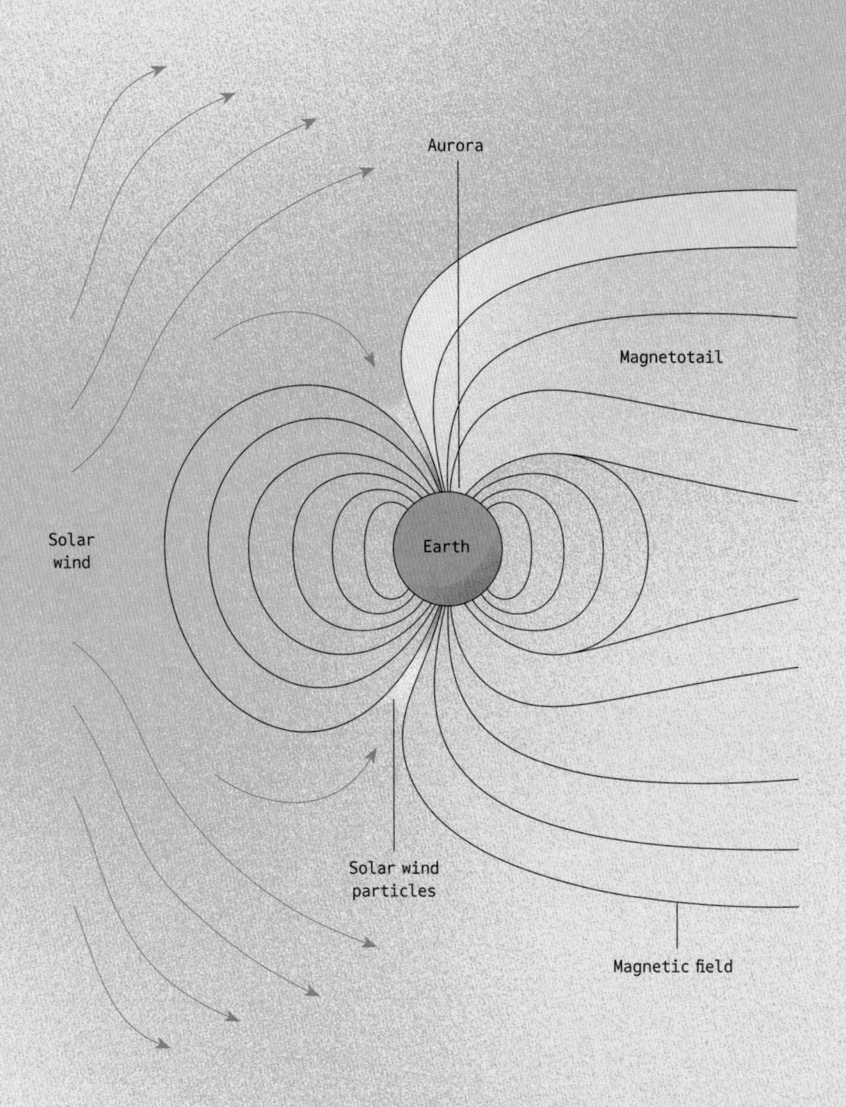

FIG. 11.1

Aurorae form when the solar wind finds its way into the upper atmosphere of the Earth, ionizing atoms and causing them to glow. If the solar wind is strong, such as after a CME, then the aurorae are visible over more of the Earth.

plunge downward until they strike one of the atoms of gas that make up the atmosphere. Carrying so much energy, these particles are able to ionize the gas, which then glows. Oxygen can glow red and green, and nitrogen will glow blue to our eyes (fig. 11.2).

The aurorae are recorded across history and across cultures, often as a fiery red glow in the sky. The most dramatic recent episode was the Carrington event in 1859, the result of a large CME striking the Earth in the era of industrialization. The world was

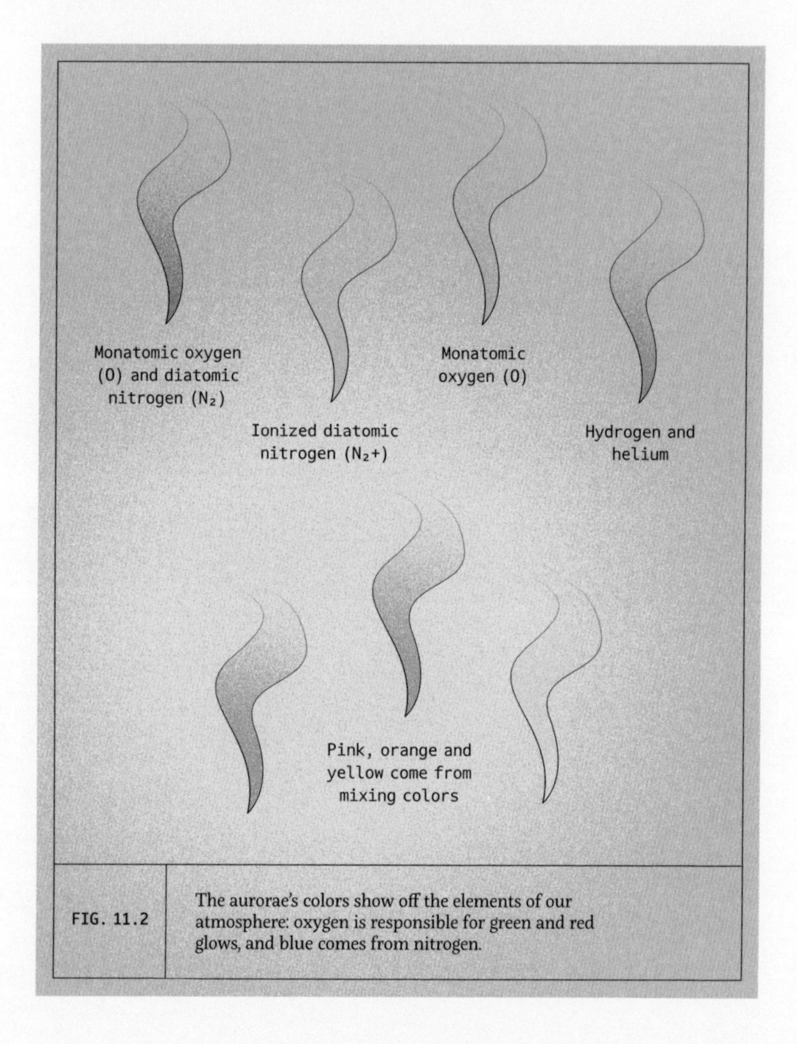

FIG. 11.2 The aurorae's colors show off the elements of our atmosphere: oxygen is responsible for green and red glows, and blue comes from nitrogen.

still operating on telegraph wires and not a modern power grid, and yet the disruption to global communications was substantial. In a modern world it's theorized that a similar event would cause "considerable disruption" to the power grid—which is a very understated way of saying that a lot of it would melt, triggering large-scale blackouts.

The Carrington event was either a strong CME or it potentially appeared stronger by having two moderate CMEs arriving more or less simultaneously. A weaker, but still extremely disruptive solar storm that caused blackouts in Quebec in 1989 may have also been two CMEs in sequence. The Carrington event doesn't seem to be particularly unique. A powerful CME in 2012 with the capacity for a Carrington-style event simply had the good manners to not be aimed at the Earth. Looking at historical records gives us an average cadence of one Carrington event hitting the Earth every 150 years or so.

In principle, any planet with a strong magnetic field in the line of a CME can experience the aurora. Even Mars, with its weak magnetic field, has aurorae, and its red landscapes might be lit by green skies of glowing oxygen when the Sun is active. More dramatically, we have fascinating imagery of aurorae on Jupiter and Saturn. Both their aurorae glow in the ultraviolet, but while Saturn's are powered by the solar wind, like the Earth's, Jupiter's occur more powerfully and more consistently than those on Earth. Jupiter has a constant source of high-energy particles much closer to it than the Sun: its volcanically active moon Io. Io dumps a lot of particles into the area immediately around Jupiter, and these resemble the ones that the solar wind provides on Earth. So rather than having aurorae that flare up and subside into nothing, Jupiter has a constantly raging storm at its poles.

in a diagram

With so many different properties that change the way we see a star, it's a good time to start working out which ones correlate with each other. One tool in particular links together many of the properties we've talked about so far—the Hertzsprung-Russell diagram, or HR diagram for short.

In this diagram, we can place the color of the star along a horizontal axis, going from bluest on the left to reddest on the right. Since we already know that the color of the surface of a star also tells us about its temperature, sometimes this axis is labeled with temperature as calculated from the color, instead of color directly, but they're coming from the same information.

On the vertical axis, we're going to place the brightness of the star. This can either be done as an absolute magnitude, or as a translation from magnitudes to how much energy the star is actually producing every second—again, two versions of the same thing.

With our diagram set up, we can take any set of stars and see where they lie in this space. In general, no matter what group of stars you look at, the vast majority of them will fall along a diagonal line, from the red, faint end of the diagram in the lower right, up toward the bluer, brighter side of the graph. This diagonal line is called the main sequence, and this is where all of those stars that are fusing hydrogen into helium in their cores lie.

The Sun can be placed on the HR diagram as easily as any other star, and we find that it sits roughly in the middle of the road both in temperature and brightness. This is partly what we mean when we say the Sun is an average star (fig. 12.1).

We mentioned earlier that the mass of a star controls how fast fusion happens in its core, and the faster fusion happens, the hotter the star must be. This main sequence is therefore also a sequence in mass. In the lower right corner are those lower mass stars that consume their hydrogen fuel slowly, and the closer to the upper left of the diagram we get, the more we are looking at high mass stars that should exhaust their fuel rapidly.

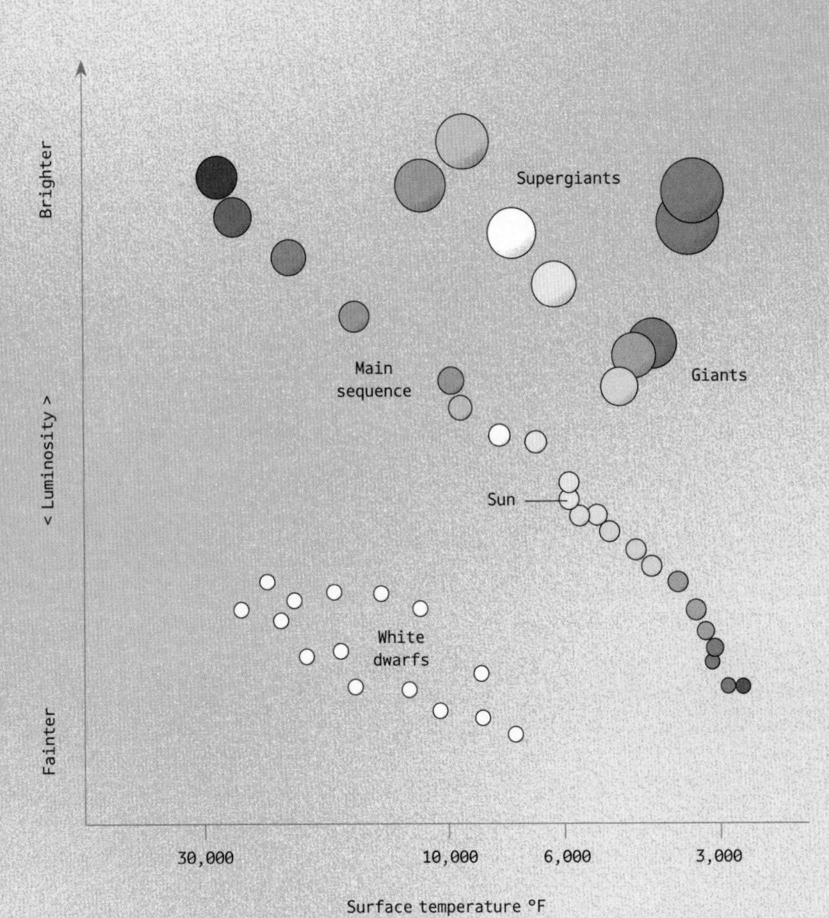

FIG. 12.1	An HR diagram allows us to place all stars in relation to each other. Most stars reside on the main sequence, which also traces the mass of stars that are able to fuse hydrogen in their cores.

While most stars are on the main sequence, there's a solid population of stars that have departed this sequence. These are stars that are now doing something different in their cores. They have exhausted all their available hydrogen, and have had to change forms in order to persist in any kind of gravitational balance. There are three broad classes of stars off the main sequence: red giant stars, which live in the bright but cool quadrant in the upper right of the diagram; brown dwarfs, which live in the extremely cold and extremely faint lower right corner; and white dwarf stars, which live in the lower left corner, for objects which are extremely faint but very hot.

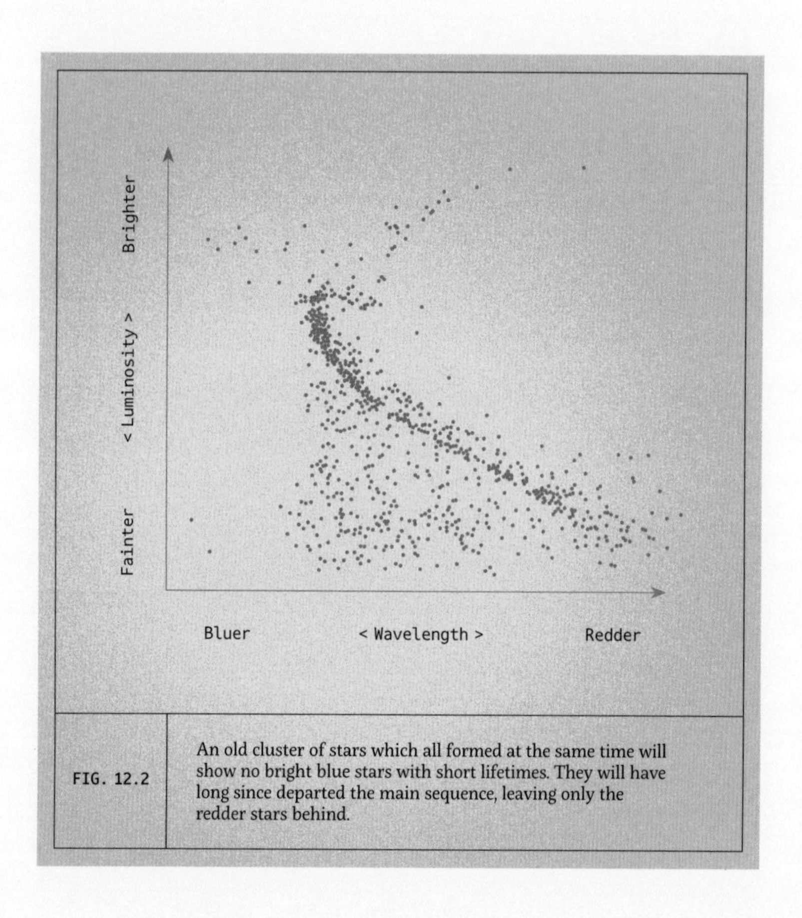

FIG. 12.2 An old cluster of stars which all formed at the same time will show no bright blue stars with short lifetimes. They will have long since departed the main sequence, leaving only the redder stars behind.

Based only on their locations on the diagram, we can immediately know some things about them. Brown dwarfs are the most straightforward. Being both cool and dim means we'd expect them to be relatively small objects. Meanwhile, to be incredibly hot but not very luminous, a white dwarf star must be exceptionally small.

If a red giant star is going to be cool but very bright, we know that it must be extremely large to produce so much light. In order to counteract the fact that each square inch of the star's surface isn't very bright, it's going to have to compensate by having a LOT of square inches.

There's another very convenient feature held within the HR diagram, which appears if you look at a population of stars that all formed at the same time, like a cluster of stars. Since we know that blue stars will tear through their available fuel much quicker than their fainter, redder counterparts, these high mass stars will also depart the main sequence earlier than the red ones. So we can use this diagram as an estimate of the age of the cluster, by looking for the upper edge of the main sequence. That upper edge will tell us, along with the masses of the stars at that edge, how long those stars have all been there. The more massive stars have already experienced their whole main sequence lifetime, while less massive stars will remain present for billions of years to come. This is called the "turn off point," and if it were placed at a 1 solar mass star's location on the diagram, that would tell us that the cluster is about ten billion years old—the expected lifetime of our Sun (fig. 12.2).

as a brown dwarf

We can also look at what happens when stars don't quite fully form. Brown dwarf stars are a misnomer from the very start. They're not true stars, nor would they appear brown to the eye. Our definition of a true star is one that can fuse hydrogen into helium in its core, and the lowest mass stars that can do this are about one-twelfth the mass of the Sun, or 80 times the mass of Jupiter. Somewhere around thirteen Jupiter masses, we start to consider the objects as just very massive planets, so this gap between thirteen and 80 Jupiter masses is filled with objects that don't quite behave like fully fledged stars, nor like massive planets.

For one thing, they do glow, faintly. They are hot enough in their interiors that they can fuse something, just not standard hydrogen. What they can fuse instead is a rare form of hydrogen, deuterium. The usual hydrogen atom is a single proton and a single electron, but deuterium is a proton, a neutron, and an electron. This lets the interior of the brown dwarf skip the first and last steps of the standard hydrogen fusion pathway. Deuterium burning just takes a deuterium nucleus and smacks another proton into it, making a less common form of helium, but nonetheless releasing energy. Deuterium fusion can only occur if our object is thirteen Jupiter masses or higher, so this is our new boundary between a very large planet and a very small quasi-star.

If we could perfectly measure the mass of every object we spot in outer space, we might be able to perfectly divide our planets from our brown dwarfs. Unfortunately, these measurements are usually a little uncertain. For some objects, we'll find that our wiggly estimate of mass is enough to say it's definitely in the brown dwarf regime, but for others it may still be questionable (fig. 13.1).

The second problem with relying on deuterium fusion as our criterion for a brown dwarf is that there usually isn't enough deuterium in the brown dwarf to keep it going for very long—only tens of millions of years at best. This short timespan of deuterium fusion means that most of the brown dwarfs we see are already out of fuel, and will spend the rest of their existence slowly cooling down, radiating their heat out into the Universe (fig. 13.2).

Iron rain and
silicate snow
2,372 Kelvin
(3,812°F)

Mostly clear skies
948 Kelvin
(1,247°F)

Cool (potentially
water) clouds
303 Kelvin (86°F)

FIG. 13.1	Brown dwarfs are only barely able to generate heat through fusion, but not through the standard pathway. They bridge the gap between very large planets and very small stars, and would be reddish purple to our eyes.

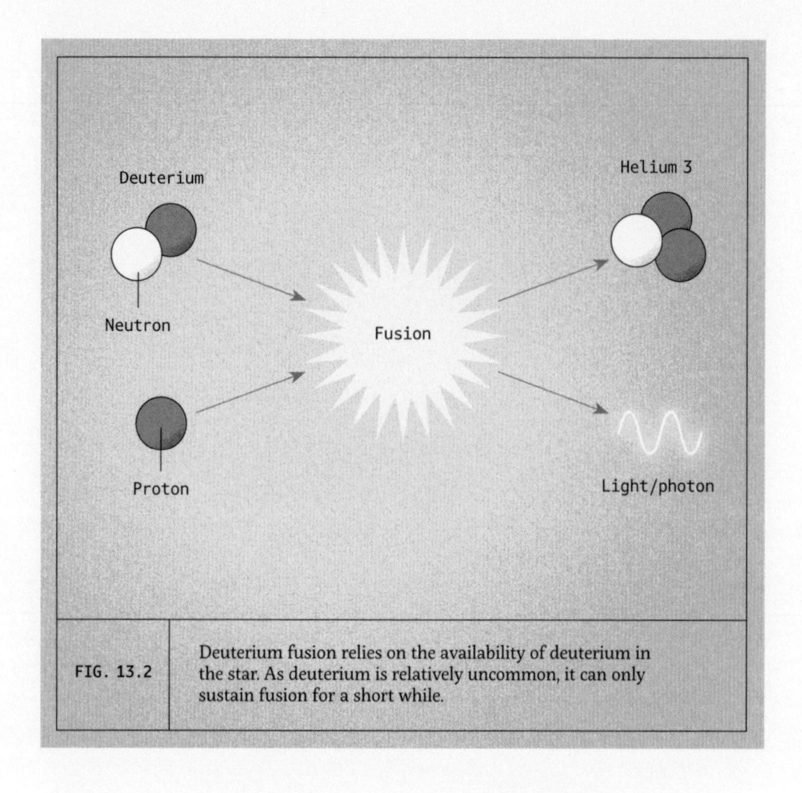

FIG. 13.2 Deuterium fusion relies on the availability of deuterium in the star. As deuterium is relatively uncommon, it can only sustain fusion for a short while.

If Jupiter is our benchmark, it glows very faintly in the infrared, with a rough temperature of about 160 Kelvin (-230°F). Brown dwarfs tend to run hotter, with the hottest around 1,255 Kelvin (1,800°F) or so. The coldest brown dwarfs so far are about 422 Kelvin (300°F), a temperature you're familiar with if you enjoy baking.

Brown dwarfs are split into three different classes: L, T, and Y, running from hottest and most massive to least massive and coldest. However, all brown dwarfs stay pretty close to the size of Jupiter. They're just much more dense than Jupiter, and this is partly why they can sustain those temperatures in the first place.

On the other hand, brown dwarfs usually have spectacularly alien weather patterns. We've spotted clouds in several brown dwarfs, often much like the banded patterns of Jupiter, but there have been other observations indicating patchy cloud formations.

The clouds in a brown dwarf, which often hold silicates and iron, rain heavier and hotter materials than Jupiter can manage. It can be sufficiently hot in the deep atmosphere of the brown dwarf (around 1,366 Kelvin/2,000°F) that iron evaporates, but when the iron reaches the outer cloud layers, it condenses back into liquid form, dropping back toward the interior. It's an iron-based precipitation cycle, and while it's nothing like rain on Earth, our kind of rain is also the closest analog that we have for this cycle, so it's often dubbed "iron rain." In other brown dwarfs, it seems to rain sand if the temperatures are right. Somewhere between 1,255 Kelvin (1,800°F) and 1,922 Kelvin (3,000°F) permits clouds of silicates—if it's colder than that, the silicate will condense out of the clouds. These sandy clouds were confirmed with direct detection for the first time in 2023, converting the inhospitable atmospheres of a brown dwarf from a very plausible scenario to a certainty.

Not all brown dwarfs have the same kinds of extreme weather; as we go to cooler temperatures, the weather shifts from sandblasting to completely cloud-free. At even cooler temperatures, clouds may return, but made of new materials; water ice may be possible.

If a human were to visit, we'd most likely see the warmest brown dwarfs as a deep red. As we visited cooler and cooler variants, the color would shift from a dark maroon down to the deepest, faintest purple. This is a trick of the human eye, which often perceives faint, deep red light as purple, and not because the brown dwarf is actually producing any blue light—it is a long way from being able to produce blue light.

as a red giant

Watching what happens when a star runs out of fuel also teaches us about its inner workings. All stars that complete their time on the main sequence will evolve off this sequence into a red giant star. Depending on their mass, this phase of their lives can be more or less complex, but there is an early phase that's shared by all stars leaving the main sequence.

The first stage of becoming a red giant is the moment when the delicate balance between fusion generating outward pressure, and gravity pressing inward, breaks down. As soon as the fusion in the core runs out of the necessary hydrogen, having fused all of it into helium, the outward pressure from fusion drops, temporarily giving gravity the upper hand. With the star unable to resist the force of gravity attempting to compress it, the innermost regions of the star are compressed down to even higher temperatures and pressures.

The inner region of the star compresses so much that the temperatures previously only found in the core are now found in a shell surrounding what is now a pile of helium leftovers from the last round of fusion. Since this shell surrounding the core was not involved in the fusion of hydrogen into helium, this layer of the star is still rich in hydrogen, and fusion can occur. This is uninventively termed "hydrogen shell fusion," and it can now generate some resistance against gravity. It generates a lot of resistance, in fact. This shell produces more energy than the core of the star did when it was on the main sequence.

Some of that extra energy from the hydrogen shell fusion is donated to the star itself, and it begins to win the fight with gravity about how big it should be. The star as a whole expands (fig. 14.1).

This expansion of the star moves it to the left on the HR diagram. The rest of the energy makes the star brighter, so it also moves vertically after the initial expansion. This curved path on the HR diagram is known as the red giant branch.

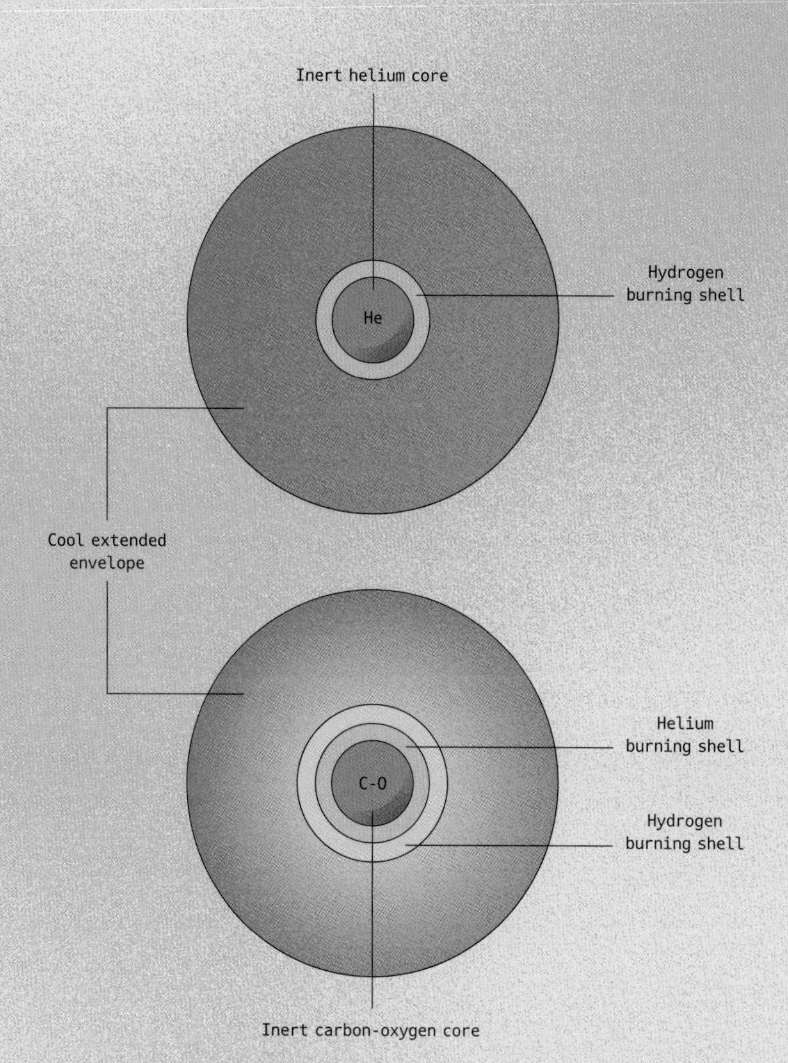

FIG. 14.1	Stars fusing hydrogen in a shell surrounding a "dead" core filled with helium are known as red giants, and are both brighter and physically larger, relative to their main sequence lives (top). A star in the asymptotic giant branch has a core filled with carbon and oxygen, a shell of helium fusion, and a further shell of hydrogen fusion, which drives the star to a brighter and redder region of the HR diagram (bottom).

Given a little more time, gravity can continue to work on the inner core, which has, in the meantime, stopped compressing. It's so dense already that the electrons are tightly packed and are resisting the gravity acting on the entire star. If we can't increase the density, then all that happens is that we increase the pressure, which increases the temperature rapidly. When the core reaches 100 million Kelvin (180 million degrees Fahrenheit), we can abruptly begin to use all that leftover helium, fusing it into carbon and oxygen.

The process of fusing helium into heavier elements requires at least three free helium nuclei, which is a reduction compared to the six hydrogen nuclei required to build hydrogen. The first stage is to fuse two helium nuclei together into a beryllium atom, and then if we can fuse one more helium nucleus into the beryllium, we get a stable form of carbon. If you then slam one more helium into that fresh carbon, you can create oxygen as a bonus. Because of the three critical helium nuclei (also known as alpha particles), this process is called the triple alpha process. (We never promised inventive names.)

The abrupt switching on of the core's triple alpha process is known as the helium flash. The helium flash produces so much energy all at once that it's able to reduce the pressure in the core and run stable helium fusion for a short time. This doesn't stop the shell of hydrogen fusion also occurring in a ring surrounding the core, so now our star has two sources of fusion pressure resisting gravity. We're creating a LOT of energy now. The surface temperature of the star increases, and so we move to the left on the HR diagram, into a region known as the horizontal branch.

Given enough time, the helium resources in the core of the star will be depleted, and another version of the entire process above happens. The now carbon and oxygen core begins to compress, and the region surrounding it becomes so hot and dense that helium fusion can begin in a shell around the core, while hydrogen fusion continues to migrate outward, churning more of the stellar interior into helium in its wake. This moves the star back upward on our HR diagram, into what's called the asymptotic giant branch.

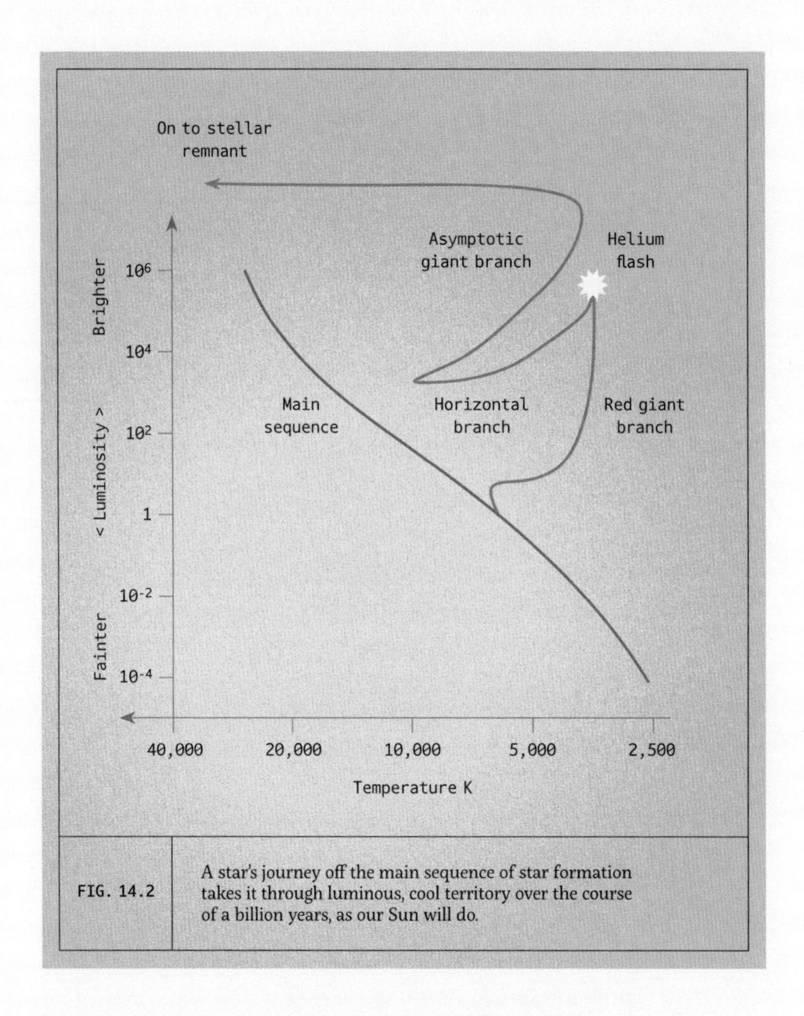

FIG. 14.2 A star's journey off the main sequence of star formation takes it through luminous, cool territory over the course of a billion years, as our Sun will do.

Helium fusion in this shell isn't particularly stable, so it has a tendency to run hot and cold as the temperature changes by tiny amounts within the star. With the erratic pressure from the interior of the star, the outer layers wind up feeling "kicks" of pressure. If the star is massive enough, it can hold on to those outer layers for a while, but a lower mass star will start to lose its grip on the outer layer of its own atmosphere (fig. 14.2).

as a planetary nebula

A strikingly beautiful way to know a star is from the temporary glory of its final stages of life: a planetary nebula. For our low mass stars (less than about 8 solar masses), the asymptotic giant branch phase is both brief—lasting at most only a million years—and also the final phase of fusion for the star. Its internal structure at this point is helium fusing in a shell surrounding the former core of the star. The core itself has accumulated enough carbon and oxygen that helium fusion has been stifled there. There's another shell of material fusing hydrogen into helium surrounding that inner helium shell fusion—and this means that the star has two sources of outward pressure. However, the helium fusion in the shell is extremely sensitive to temperature and therefore pressure, so with mild changes in the star's pressure balance, the rate of helium fusion can change a lot.

If the rate of fusion changes, then the outward pressure and the temperature generated in the core of the star are also changing, and it's the outer layers that will feel this most intensely. These bursts of increased fusion rate from the helium shell are called thermal pulses, and they have the impact of repeatedly kicking the outer layers of the star outward. As this continues to happen, these outer layers become gravitationally untied from the inner regions of the star, where hydrogen and helium fusion are happening, and the star loses mass. Each time there's another thermal pulse, there's an additional kick to the new outermost layers, and more of the star is lost to interstellar space.

Given enough time, the helium fusion in the core kicks so much of the rest of the star away with these abrupt pulses of energy that, in its final throes, it pushes away even the innermost source of fusion, the helium shell, and leaves the helium-exhausted core exposed to the deep dark of interstellar space (fig. 15.1).

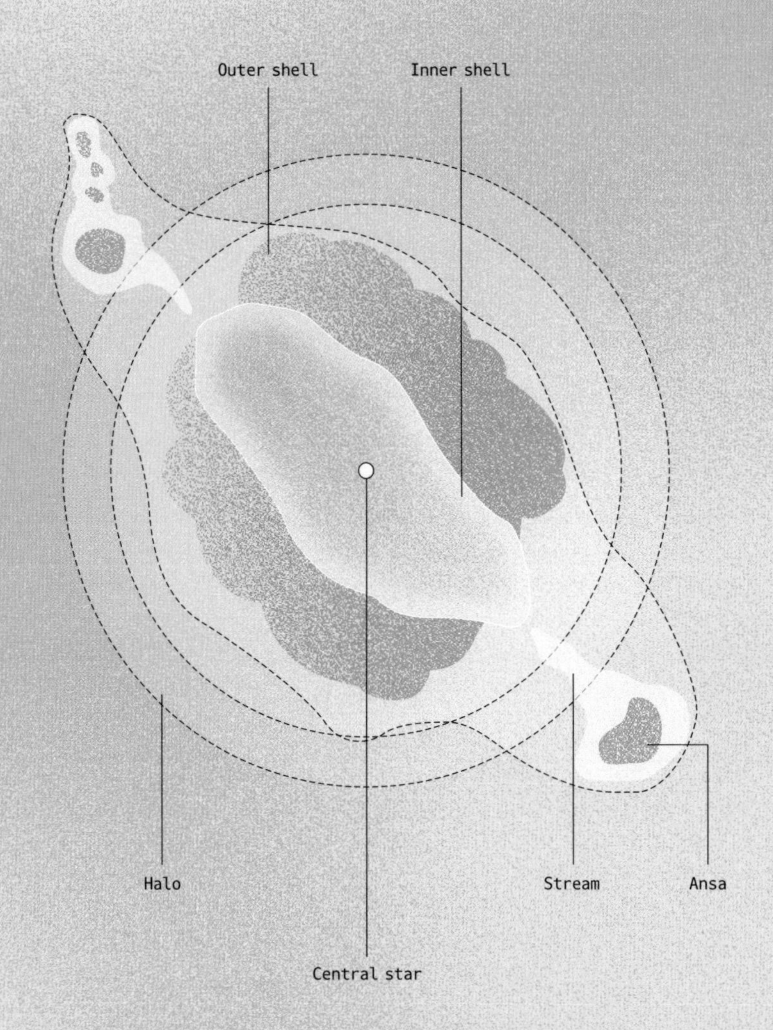

FIG. 15.1	Stars create beautiful and temporary planetary nebulae in the final stages of being able to fuse elements within them, shedding the outer layers of the star into a nebula much larger than our solar system is now.

The former stellar atmosphere, in its new journey away from the remains of the star, is a spectacular, though temporary, sight. Heated to very high temperatures by what remains of the interior of the star, the former atmosphere is now distributed over several light years, and glowing.

All planetary nebulae are young; they typically last only tens of thousands of years. After that, the gas that makes up the nebula is simply too faint and thin to be seen, and will continue to drift in the space between the stars.

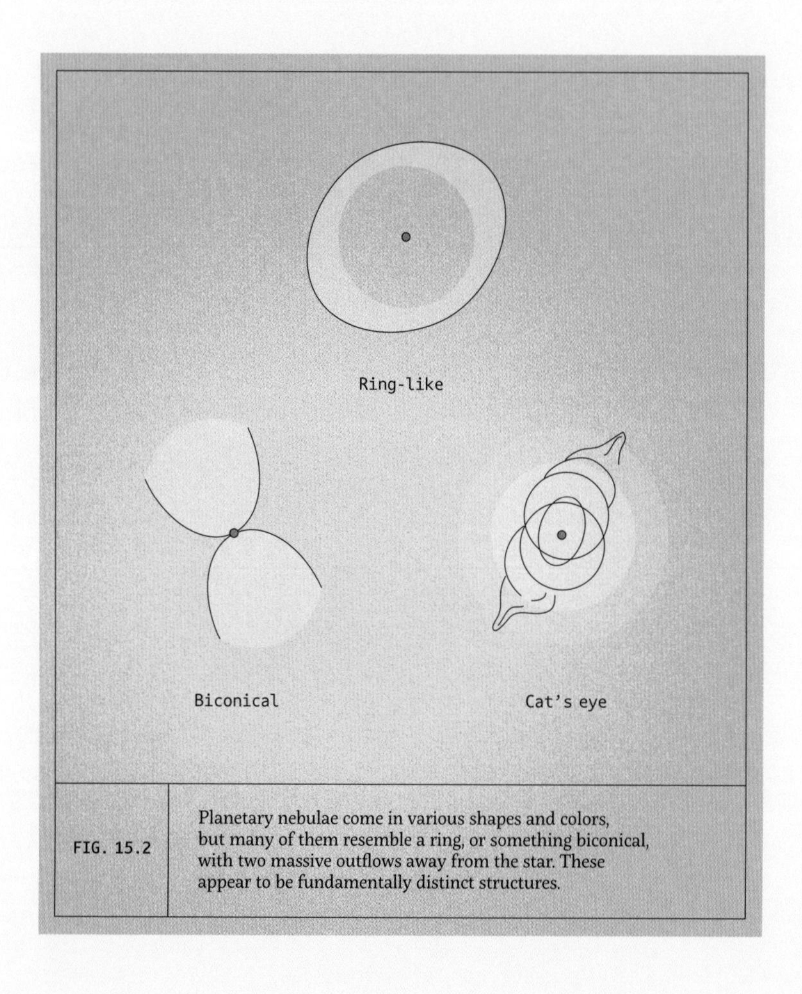

FIG. 15.2 Planetary nebulae come in various shapes and colors, but many of them resemble a ring, or something biconical, with two massive outflows away from the star. These appear to be fundamentally distinct structures.

Planetary nebulae are dramatic, in part because of their vivid colors, and in part because no two of them look exactly alike. Most of them, however, can be classified into a few general shapes—ring-like, elliptical (like the Cat's Eye Nebula), and biconical—though of course there are always nebulae that don't fit into any of the above (fig. 15.2).

One of the many puzzles about planetary nebulae is how one process—losing the outer layers of the star—can result in such visually diverse nebulae. One possibility is that we're seeing the same general shape of object at very different angles. A biconical outflow might look circular if you saw it from the top down, or elliptical if the angle wasn't quite right. However, as our ability to observe these nebulae has gotten better with time, we've been able to start creating 3D models of these structures, and not all of them are hiding biconical shapes.

The Ring Nebula, one of the most photogenic planetary nebulae due to its closeness to us, looks circular at first glance, but current observations tell us it's a mixture of not-quite-round structures. Instead, it seems to be a donut of glowing nitrogen and oxygen, and then a longer oval of superheated helium that we happen to be seeing nose-on. These structures combined, plus the way we're looking at it, result in something that looks reasonably circular. A different viewing angle might have given us another perspective, but no angle would make it look like the biconical outflows of other planetary nebulae, such as NGC 6302.

So that takes us back to one physical process truly creating different shapes of planetary nebula, in which case there must be something about the star or its system that can change the shape of the nebula it creates. One suggestion is that the mass of the star may be important. Stars with the lowest mass seem to produce the roundest, brightest, and most symmetrical nebulae, with the higher mass stars producing fainter and more asymmetrical systems. The high mass systems are also more likely to produce the biconical shapes. Since binary stars are also more common the more massive the star gets, it's also possible that the presence of a second star allows for more complex and asymmetrical shapes to appear.

PLATE 4

Colorful rings of a former star

The Ring Nebula is a famous planetary nebula. In this wider view, faint outer structures from the earliest layers of the star to be lost are visible as a vaguely floral series of red shells. Inside the main body of the nebula, which is about a light year across, ionized gas glows brightly.

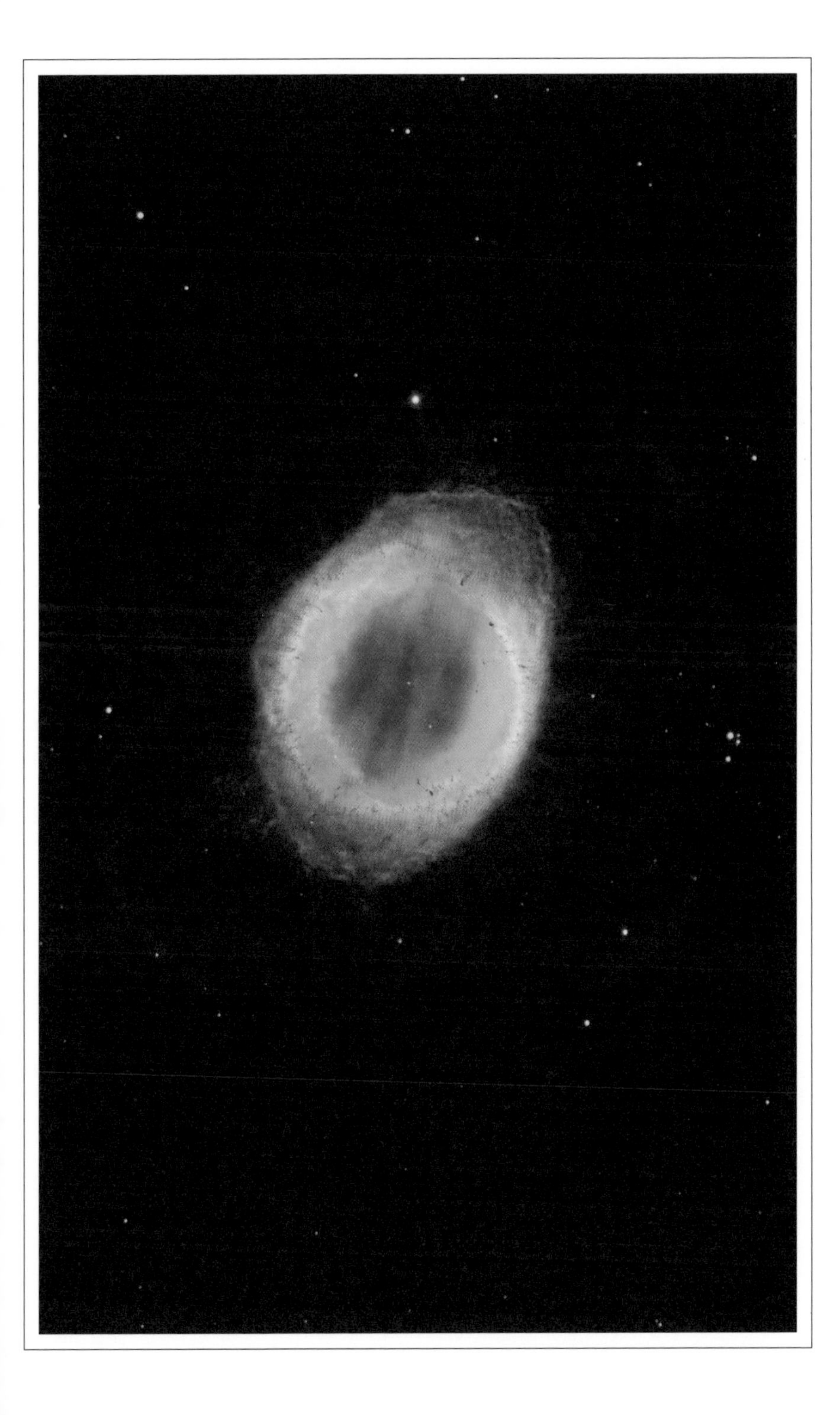

as a white dwarf

To dive even deeper, we can examine what's left behind after the brilliant glow of its planetary nebula fades. What remains is the superheated former core of the star, filled with carbon and oxygen, the end products of helium fusion. This core is tremendously hot, but too cool to form any new elements, and not massive enough to compress itself to a higher density.

Once the core of the star is exposed, it's now known as a white dwarf, and is officially in a class of objects known as stellar remnants. These are the leftover objects after a star finishes all the fusion it can. All stars less than about eight times the mass of the Sun will end up as a white dwarf. These profoundly dense objects often contain 60 percent of the mass of the Sun, packed into an object around the same size as the Earth.

With no further source of fusion, the white dwarf is done generating new heat. All it can do is slowly radiate the heat it has already built up out into the cosmos. In principle, white dwarfs should eventually cool all the way down to end up just as cold as the space that surrounds them. In practice, current models suggest that would take some trillions of years, which is substantially longer than the Universe has been around (approximately 13.8 billion years). We do not expect that any white dwarfs have had the necessary time to cool into what would be called a black dwarf (fig. 16.1).

For a star like the Sun, models predict that about half its mass would be lost in the red giant and planetary nebulae phases, leaving behind a white dwarf of about 0.54 solar masses. More massive stars will leave behind more massive white dwarfs, and do so sooner, since they will end their main sequence lifespans faster than those stars which are smaller and cooler. This is why the average white dwarf is a little more massive than the Sun's predicted end state, at 60 percent of the Sun's current mass. White dwarfs more than 65 percent of the Sun's mass are relatively uncommon, and by the time they reach the Sun's mass, we're in the rarest 1 percent of white dwarfs.

Sirius B, at 4.31×10^{30} lb (1.96×10^{30} kg)
and 3,629 miles (5,840 km) radius

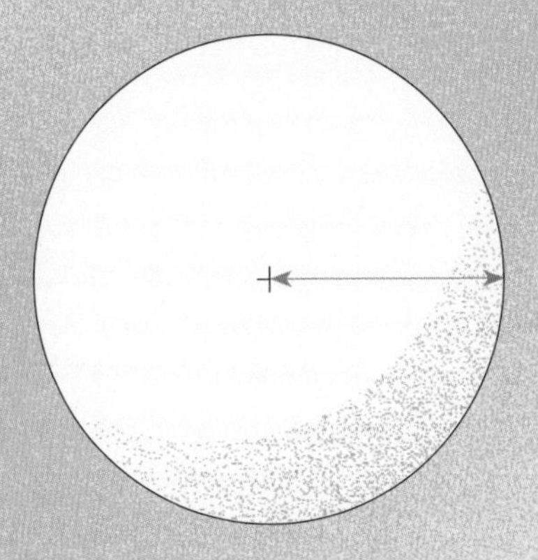

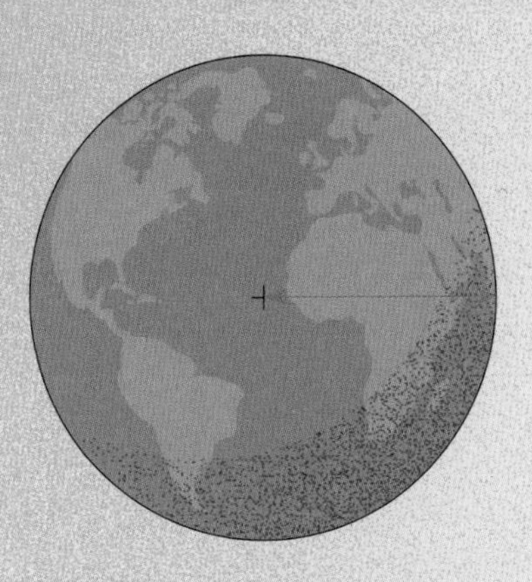

Earth at 1.31×10^{25} lb (5.97×10^{24} kg)
and 3,963 miles (6,378 km) radius

FIG. 16.1	White dwarfs are stellar remnants so dense that the mass of the Sun can be packed down in a volume smaller than the Earth in some cases, as is true of Sirius B.

In a stroke of purely random luck, the nearest white dwarf to Earth is 0.98 times the mass of the Sun, in the extremely rare echelon of relatively massive stellar remnants. And it's easy to find in the sky, since it's a companion to Sirius, though not easy to spot directly. Sirius A, as the brightest of the companion stars is called, is orbited by Sirius B, a 1 solar mass white dwarf, slightly smaller

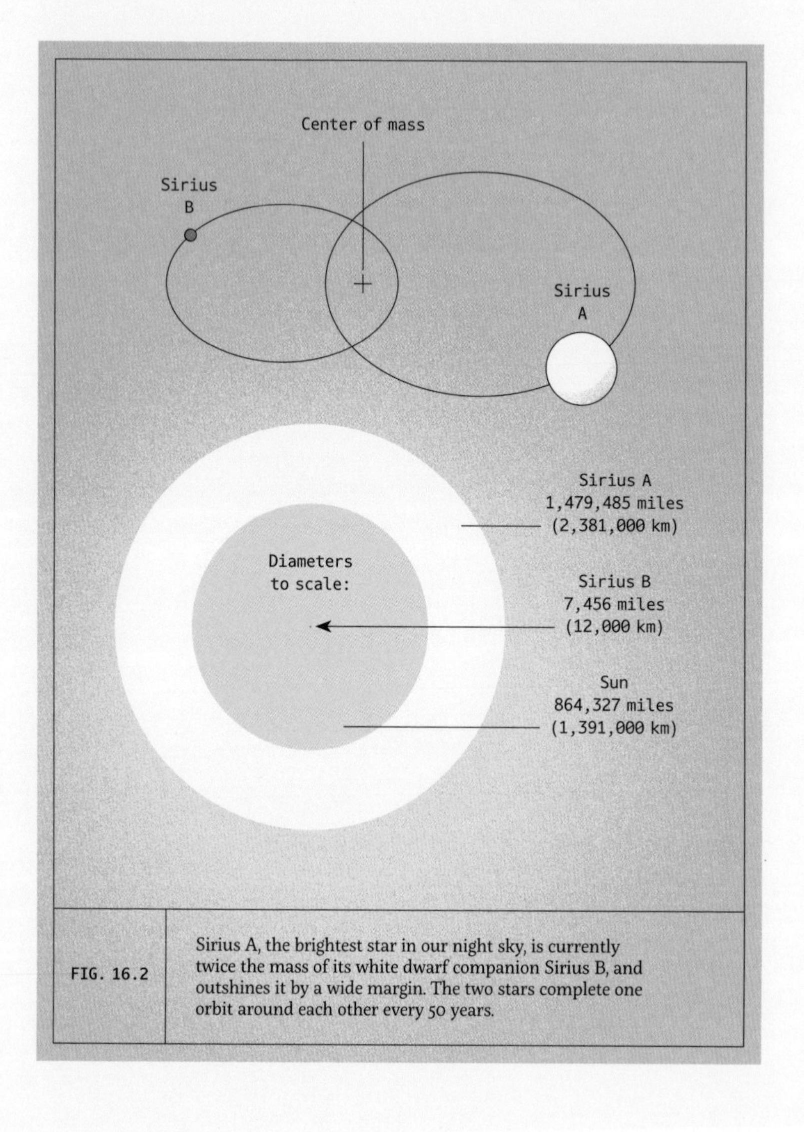

FIG. 16.2 Sirius A, the brightest star in our night sky, is currently twice the mass of its white dwarf companion Sirius B, and outshines it by a wide margin. The two stars complete one orbit around each other every 50 years.

than the Earth. These two stars complete their circuit around each other once every 50 years. Sirius A is currently the brightest star in our night sky, with a mass of 2.02 solar masses, but the star that created Sirius B would likely have been brighter—estimates put that star at somewhere around 5 solar masses. This also tells us that while Sirius A will eventually become a white dwarf at the end of its own life, it will create a less massive white dwarf than Sirius B. We're observing these two stars at a unique time in their evolution (fig. 16.2).

Without a source of fusion in their cores, white dwarf stars need another way of resisting collapse due to the constant inward pressure from gravity, and they have found it in electron degeneracy pressure. This arises when material is so profoundly compressed that electrons begin to be squeezed together. In a white dwarf, the material is still fully ionized, which means that the electrons are not bound to a specific atomic nucleus. However, there are still an equal number of electrons floating around as there are protons, and with the white dwarf compressed down to a sufficiently small size, there's only so much volume these electrons can move around in. As the volume compresses further, electrons wind up needing to share space more and more, and since they are all electrically negatively charged, they resist this need to share space. This resistance will grow as the density of material increases, and there is a point at which the electron degeneracy pressure is so high that gravity is no longer able to compress the material any further. The white dwarf is held up against gravity purely by electrons not wishing to share space.

This is a stable configuration. Unless something changes the mass of the white dwarf, electron degeneracy will not run out of fuel, and neither will anything modify the inward pressure of gravity, and so the two forces will balance each other. Electron degeneracy pressure only cares about the density of the plasma, so the gradual cooling of the white dwarf won't change this balance.

as a repeating explosion

In watching the stars, we can observe some of them repeatedly burst into light. To recreate this event, we must begin with a white dwarf, the exhausted former core of a reasonably low mass star. White dwarfs are, in isolation, unable to do anything aside from simply radiate their heat out into the surrounding space, and slowly cool down.

However, if the white dwarf has a nearby companion star of lower mass (therefore with a longer fusion lifetime), then interesting things can happen. We need the companion star to be of lower mass because we would like it to still be a star, and not another white dwarf. And the reason we'd like it to still be a star is that to create this repeated explosion in our skies, the white dwarf needs to steal the atmosphere of its neighbor, gravitationally siphoning hydrogen away from it.

If the companion star is still on the main sequence, then the two objects need to be quite close to each other in their orbits. If the binary companion is a red giant, with its large outer atmosphere, they can get away with being further apart. In any case, an outer layer of the star needs to drift close enough to the white dwarf that gravity transfers control of that gas to the white dwarf, instead of it staying attached to the star.

White dwarfs aren't particularly great at landing new material on their surfaces, so what tends to happen first is that the white dwarf winds up with a swirling disk of material surrounding it, which gradually will accumulate at the surface of the white dwarf. Accumulating material like this on top of a white dwarf is temporarily fine, but once the material begins to pile up, it will have explosive consequences (fig. 17.1).

Because of the density of our white dwarf, gravity at the surface is prodigious, and so the stolen material will be compressed down into the surface. Given that the white dwarf is helping itself to the plasma from the outer layers of its companion star, this plasma is almost entirely hydrogen. And, as we learned from the cores of stars, if we compress hydrogen down enough, we can heat it while making it denser. These are the requirements to begin fusion.

FIG. 17.1	When a white dwarf has a nearby stellar companion, the star can have some of its outer layers continuously siphoned off, and these accumulate on the surface of the white dwarf.

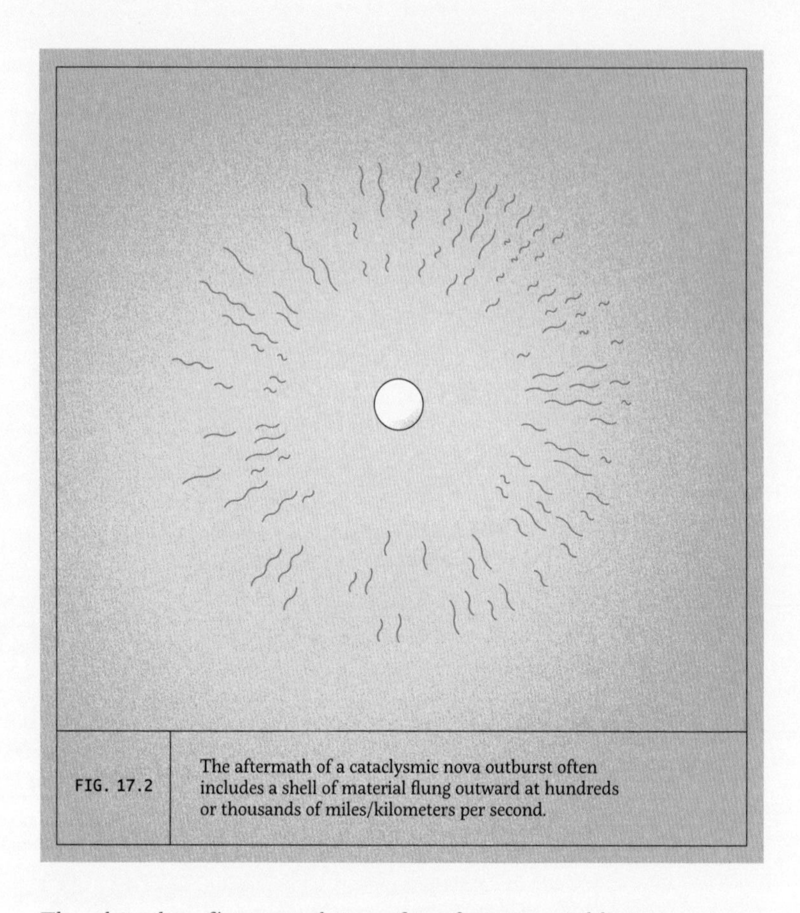

FIG. 17.2 The aftermath of a cataclysmic nova outburst often includes a shell of material flung outward at hundreds or thousands of miles/kilometers per second.

The white dwarf's accumulation of gas from its neighbor can trigger a brief, vigorous burst of hydrogen fusion on its surface. This outburst of fusion does two things. First, it creates quite a lot of light, which means that these events are easily observable. We have dubbed them "novae," from the Latin for "new." Quite often the white dwarfs that undergo such a nova explosion are intrinsically quite faint, but as the nova increases their brightness by up to nineteen magnitudes (nearly 40,000,000 times brighter), the nova can sometimes produce a "new star" in the sky, which will fade away as the fusion episode ends and the white dwarf fades back down to its original, fainter state.

Gravity will compress and heat the deepest material, closest to the pre-existing surface of the white dwarf, the most intensely. This, in turn, means that it's the innermost layer of accumulated hydrogen that will meet the criteria for fusion first, while the outermost layers—though still extremely compressed—will not meet the same criteria at the same time. This density layering leads to the second major phenomenon of a nova (fig. 17.2).

Not all of the accumulated hydrogen is able to fuse. In fact, only a small fraction of it does. But the abrupt switch to fusion in the deeper layers of the hydrogen means that there's a lot of outward pressure all of a sudden. So the innermost layers produce a rapid burst of pressure, which explosively flings the outer layers away from the white dwarf, creating a shell of material. This material can be observed with modern telescopes, and is seen to be moving at speeds more than 620 miles (1,000 kilometers) per second away from the white dwarf.

Perhaps surprisingly, this whole process does exactly zero damage to the white dwarf. As far as it's concerned, it might as well have sneezed. Nothing about the circumstances of the binary star, its orbit, or the mass of the white dwarf itself has changed. Functionally, all that has been achieved through this explosion is that the system of two stars has been reset to where it was before hydrogen began accumulating. Which, in turn, means that there's absolutely nothing stopping it from recurring. Most novae are part of repeating systems, though the timing between explosions depends on how close the two stars are, and how rapidly material is transferred from one to the other.

as a red supergiant

At the other extreme, we can learn about the onion-like end of life of massive stars. Massive stars—those that clock in at 8 solar masses or above—have short lifetimes on the main sequence, and begin to end their lives after only about 50 million years. If they're more massive than that, the "normal" hydrogen fusion lifespan is even shorter, at about 1 million years for a star 40 times the mass of the Sun.

A massive star goes through the same first stages of transforming into a red giant as its low mass counterparts. It first exhausts the hydrogen reservoir in its core, and begins to fuse hydrogen into helium in a shell surrounding the core, until the temperatures in the core can rise to the level needed to begin fusing helium. Unlike a lower mass star, helium fusion is less disruptive to the balance between gravity and pressure, so there's no helium "flash" in a more massive star, and it begins smoothly. Helium fusion will eventually be exhausted in the core, and a shell of helium fusion will add to the shell of hydrogen fusion, leaving behind a carbon- and oxygen-rich core. In a lower mass star, this is where the story would end.

But in a high mass star, there's so much material present that gravity is capable of compressing this carbon and oxygen core even more, beginning the fusion of heavier elements. First up is carbon fusion. In this process, two carbon atoms slam into each other hard enough that they fuse, which typically creates either neon and a helium nucleus, or sodium and a hydrogen nucleus. The sodium can absorb the proton created when the sodium is formed, which then turns it into neon, plus a helium nucleus (fig. 18.1).

The end result of this carbon-fueled process is a plasma stew of neon, sodium, and the oxygen left over from the helium fusion, but largely it will be neon and oxygen. The next stage fuses the neon. Neon can either absorb a high energy photon and split apart into oxygen and helium, or it can be struck by a helium nucleus and turn into magnesium. However, when true neon fusion begins, it takes two neon nuclei and fuses them into one oxygen and one magnesium atom. This, in turn, builds up a new layer of our plasma onion, made up of magnesium and oxygen.

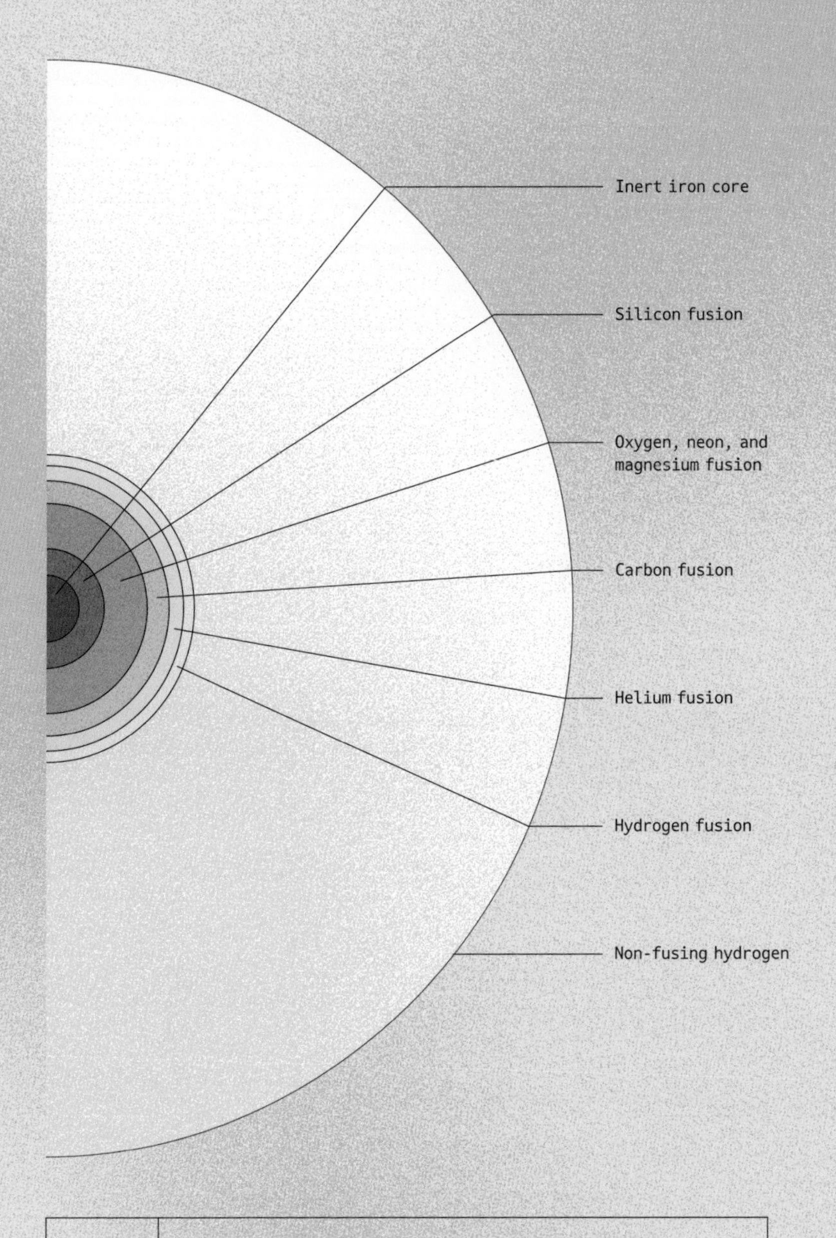

FIG. 18.1	A high mass star can continue to fuse more elements beyond helium, and so eventually builds up a large collection of shells of fusion.

Next to fuse is oxygen. Two oxygen atoms collide to create silicon and helium, or sulfur and a neutron, leaving us with a core filled with silicon and sulfur, an oxygen fusing shell, a neon fusing shell, a carbon fusing shell, a helium fusing shell, and a hydrogen fusing shell. There's only one more layer left to build, and it's short-lived.

While all of this has been ongoing, the star has been losing hold of its outer layers, and expanding to an enormous size. Stars in this condition are often hundreds of times the size of the Sun, and tens of thousands of times brighter. In our HR diagram, they live in a similar space to the standard red giant star, but at higher luminosities to reflect their larger size.

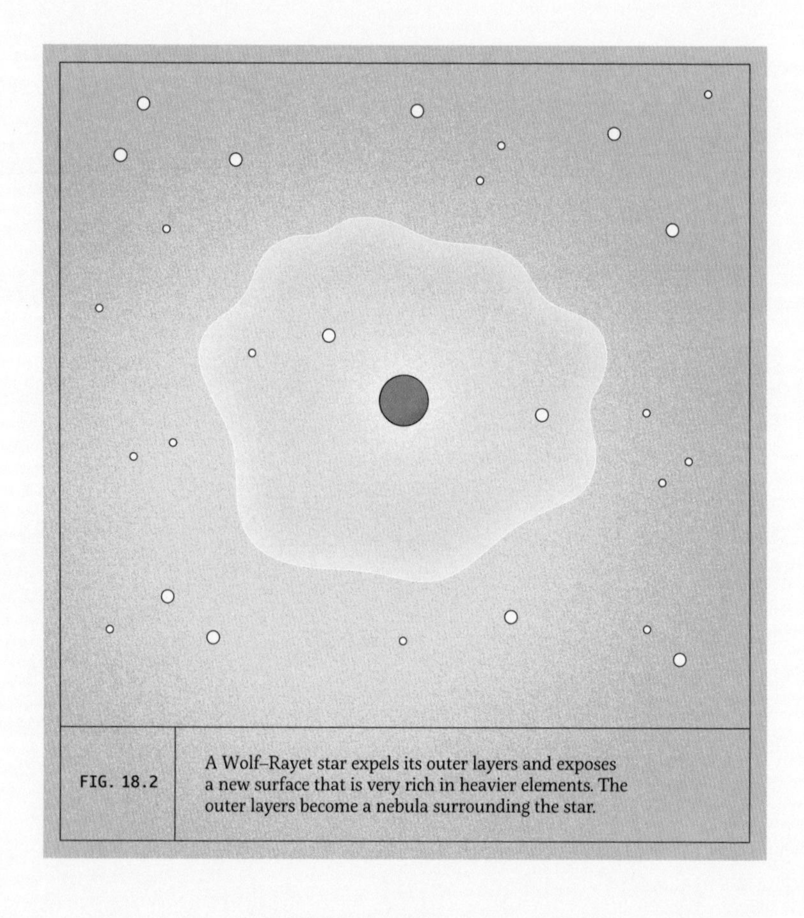

FIG. 18.2 A Wolf–Rayet star expels its outer layers and exposes a new surface that is very rich in heavier elements. The outer layers become a nebula surrounding the star.

The final stage of fusion is a cascade of element formation, which begins with the fusion of silicon. The temperatures and pressures are so great that the elements present can, as with neon before, come apart at the seams simply from absorbing a high energy beam of light, which produces helium nuclei. This is convenient, as we need that helium nucleus in order to strike our silicon nucleus. Silicon plus helium gives us a sulfur nucleus. Add another helium nucleus to sulfur, and we get argon. Add one more to get calcium. Another gives us titanium. Another gives chromium, another gives an unstable form of iron, then on to an unstable form of nickel, and then to stable iron.

And then it stops.

Iron is the final element that can be fused in the core of a star, and for a brief, shining moment, the star is an elaborate onion of plasma, forming many new elements simultaneously, and prodigiously luminous. This is the second to last moment of a massive star's lifetime.

It is thought that there is one more possible pathway to reach the end of a star's life: a Wolf–Rayet star. This pathway seems only to be taken if the star formed with more than 40 times the mass of the Sun. These stars are rare, and have main sequence lifetimes of only half a million years, but are distinct because they lose their outer layers of hydrogen entirely as they continue onward with fusion of heavier elements. The exposed surface of the star is thus rich in helium and other heavier elements, with the hydrogen lost to the region surrounding it. They are often seen with dramatic clouds around them, and are expected to be a short-lived phase in a short-lived stellar lifetime (fig. 18.2).

PLATE 5

A massive star sheds its skin

A Wolf–Rayet star (WR 124) in full bloom. This is a massive star at the end of its lifetime. The bright cloud that surrounds the central star is the remains of what was once the outer layers of the star's atmosphere.

as Betelgeuse

We need look no further than Betelgeuse, the bright red star in the constellation of Orion, for one of the best examples of a red supergiant star in our night skies. Visibly red even to the unassisted eye, it's very bright, made more impressive by the fact that current estimates place it around 550 light years away from Earth. Betelgeuse is one of only a few stars that can be seen in more detail than an infinitely tiny point. Telescopes from the ground have been able to observe hot patches on its surface, some 200 Kelvin (360°F) hotter than the surrounding atmosphere.

In spite of these unique vantage points, Betelgeuse comes with its own challenges. Our ability to see a supergiant star in such detail comes with the side effect of making the task of explaining it much more complicated. Even the distance estimate is uncertain. For such a bright star, we still have 15 percent error, which makes concrete statements about its brightness, size, and other properties difficult.

Challenges notwithstanding, models have given us estimates of what Betelgeuse is doing now. These suggest that it started its life as a star somewhere between 18 and 21 solar masses, which would indicate that it spent about 6 million years on the main sequence before transitioning into a red giant state. In this state it seems to have already lost a good amount of mass—its present-day mass is estimated between 16.5 and 19 solar masses, so it's very likely lost as much mass as currently exists in our entire Sun. Older estimates of the radius of Betelgeuse placed it somewhere between 500 times and 1,100 times the Sun's radius. A newer model places the star right in the middle of this range, at around 764 times the Sun's current radius. This translates to about 3.5 astronomical units, a distance which in our solar system would engulf all the rocky planets and the vast majority of the asteroid belt (fig. 19.1).

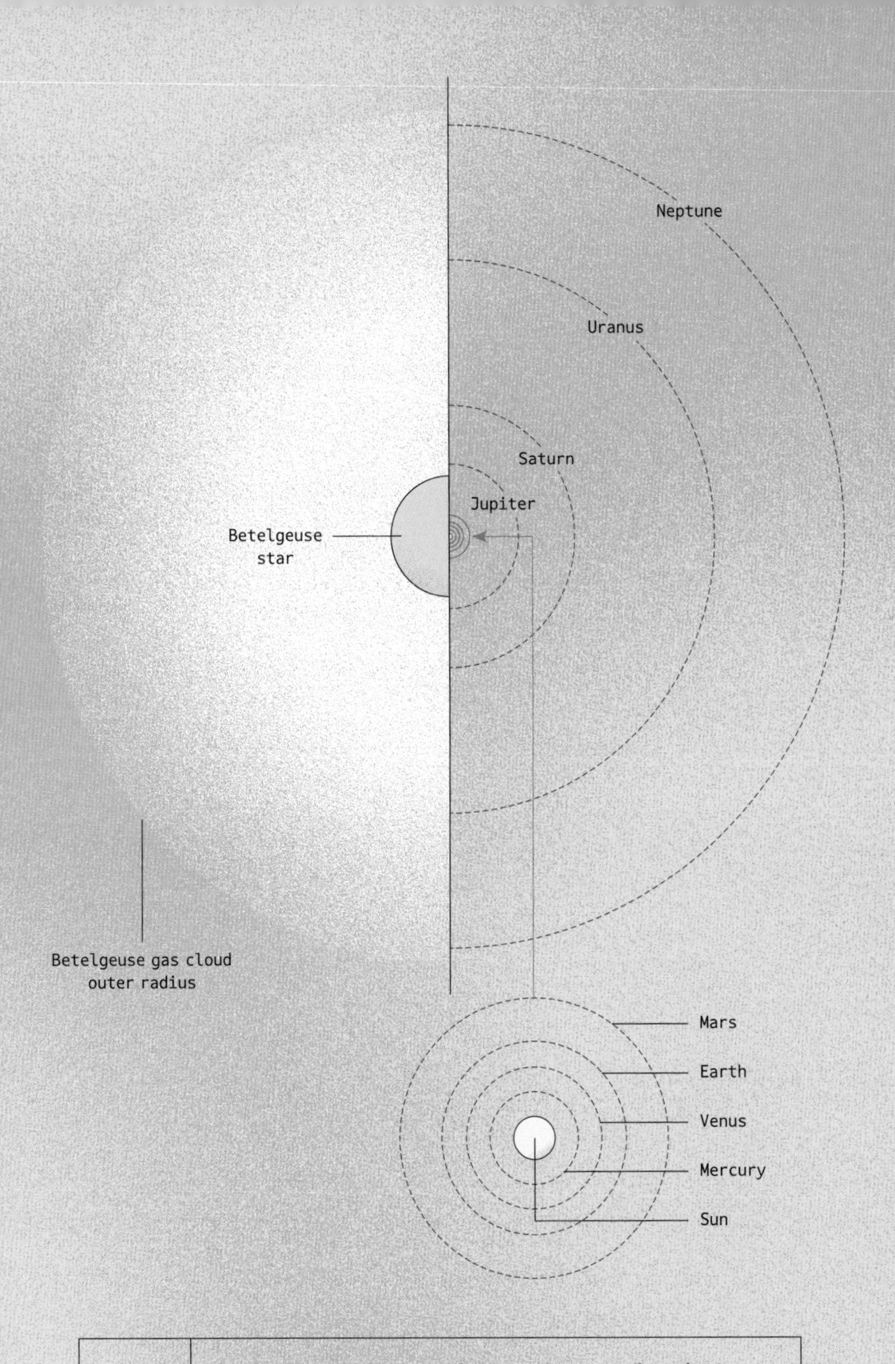

FIG. 19.1

Betelgeuse, one of the closer red supergiant stars, is so large that it would engulf all of the rocky planets in our solar system, with the roiling surface reaching to the outer edge of the asteroid belt.

Not helping our attempts to model Betelgeuse is that the star itself is variable. Its brightness varies, going from brighter to fainter every 185 days or so, a second cadence of 415 days, and a third cadence of 2,160 days (just under six years), observed over decades. This makes for a star that is almost always one of the brightest ten stars in the night sky, but if you look carefully, its brightness goes through a complicated pattern. However, this variability is a piece of the evidence that tells us the star is definitely in some form of supergiant phase and not some other weird kind of system. Models also suggest that Betelgeuse is in the early stages of fusing helium in its core—it hasn't yet accumulated enough carbon and oxygen in its core that it has begun any helium shell fusion.

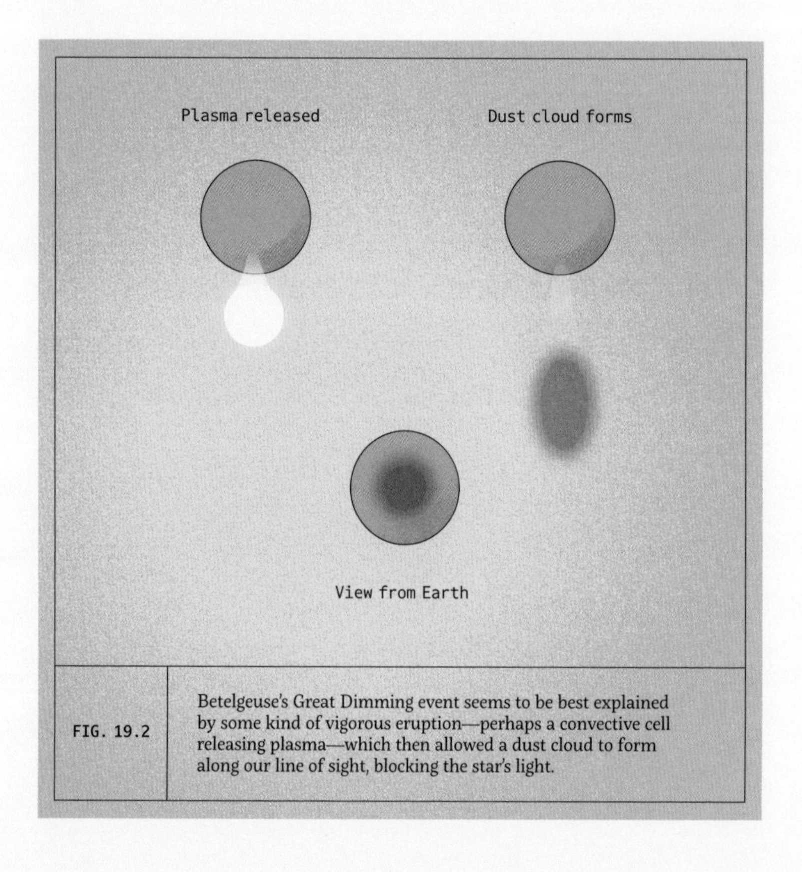

FIG. 19.2 Betelgeuse's Great Dimming event seems to be best explained by some kind of vigorous eruption—perhaps a convective cell releasing plasma—which then allowed a dust cloud to form along our line of sight, blocking the star's light.

Even with all these puzzles, Betelgeuse surprised everyone in late 2019 by dropping more than one magnitude in brightness—a factor of more than 2.5 drop in luminosity. This drop was apparent to the unaided eye, and because Betelgeuse is normally such a recognizable star, it gained a considerable amount of attention. Not least, scientists were trying to figure out what was happening with the surface of this star.

The first thing determined was that Betelgeuse was not about to undergo any kind of dramatic end of life event. However, it was fainter than expected, even for any of the known brightness variations. The 415-day drop—which it did align with—usually lowers the brightness by 0.3–0.5 magnitudes, which is only able to account for half the drop in brightness. Dubbed the Great Dimming event, it persisted over several months, gradually dropping in brightness before Betelgeuse returned to normal levels, where the star has remained since then (fig. 19.2).

Betelgeuse was a popular target for telescopes during the few months when it was dimming, and observations taken with the Very Large Telescope in Chile revealed that the southern half of the star had faded relative to the northern half, sitting about ten times fainter than usual. This summarily broke any kind of symmetrical behavior explaining the dimming of the star.

Other telescopes supported the idea that there had been a dust cloud blocking the light from the star. In all likelihood, it wasn't one that pre-existed and just happened to drift in between us and Betelgeuse, but was potentially something like a bright burst of plasma ejected from the star (which matched observations from the Hubble telescope). This would have cooled as it left the surface and formed a large cloud of dust which, in turn, blocked the light from the star. From our perspective, the star brightened and then faded, as the cloud of dust was given the means to form and then blocked out some of the light from the star. As the cloud dispersed, Betelgeuse could return to its former brightness

[TWENTY] Know a star...

by an explosive end

A particularly impressive way to know a star is by witnessing one of the cosmos' most luminous events, the final stage of life for a high mass star: a core collapse supernova. In order to build our core collapse supernova, we need to pick up where we left off with our high mass stars.

At stellar masses over eight times the mass of the Sun, the star will have built up shells of fusion within its interior. Having begun the silicon burning process, it is going to accumulate iron in its core, as the unstable form of nickel decays into iron. Silicon burning is a rapid process, and will complete in about 24 hours for a star 25 times the mass of the Sun. Unlike the carbon core which is left behind relatively unchanged in a low mass star, the iron core will not remain as a stable object.

The gravitational pressure now is unresisted, pressing inward on the iron core and heating it to prodigious temperatures. These temperatures mean that there are extremely high-energy photons bouncing around. As soon as an iron atom encounters one of these, it falls apart, all the way back down to helium. This process is known as photodisintegration, and is the closest the cosmos comes to a disintegration ray. The pressure is so high that the electron degeneracy that supported our white dwarf stellar remnants is overcome, and electrons are shoved into protons to make more room. This creates a large pile of neutrons and an equally large pile of neutrinos. Neutrinos are tiny fundamental particles which, under normal circumstances, only barely interact with matter. But our supernova is nowhere near normal circumstances, and these neutrinos will become important.

The destruction of the iron nuclei and the disintegration into a pile of neutrons is extremely rapid. This is our core collapse, and it takes about a quarter of a second (fig. 20.1).

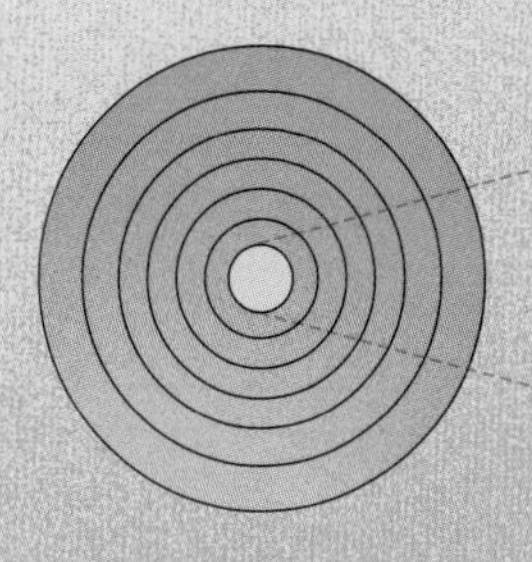

Just before supernova, the star is made of many layers of elements

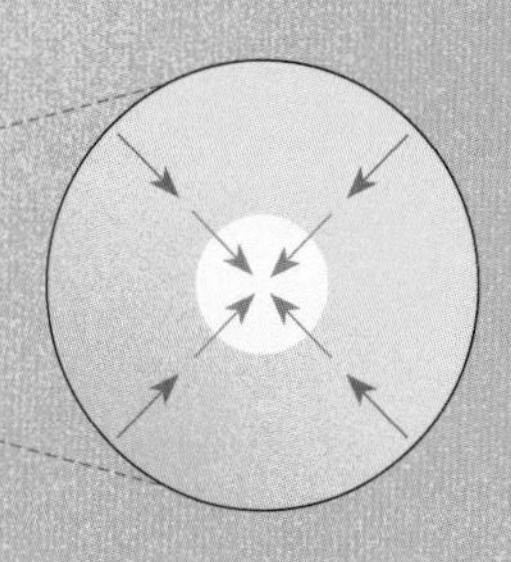

The core collapses under its own mass as iron disintegrates

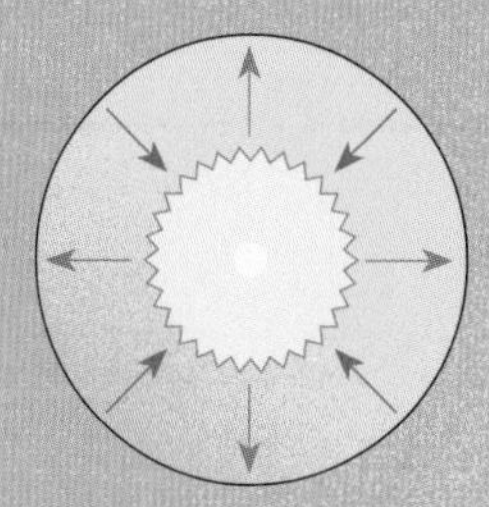

Briefly, the neutrons resist further collapse, starting a shockwave

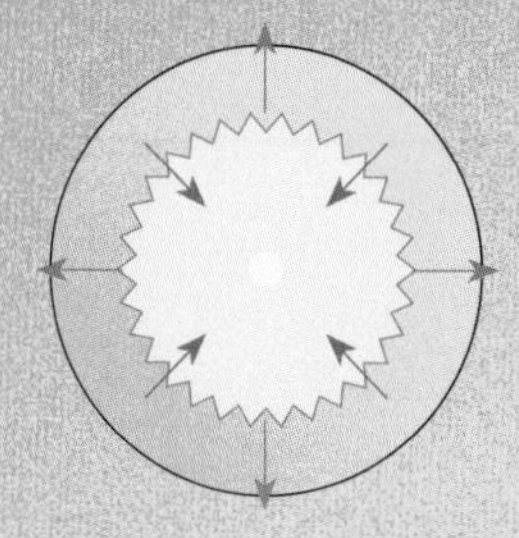

Neutrinos, created with the neutrons, push the shock front outward

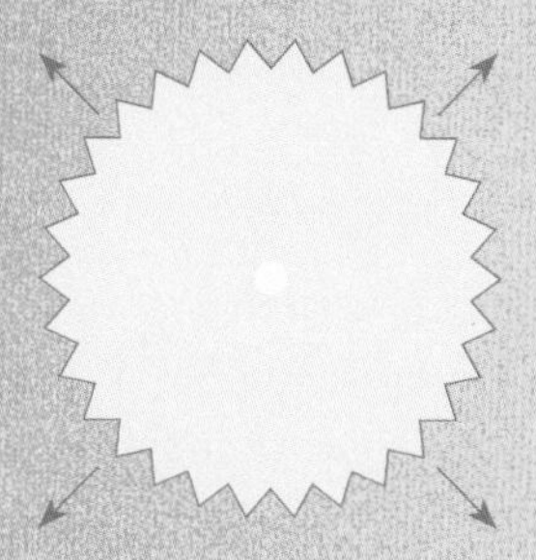

The explosion proceeds outward as a supernova

FIG. 20.1	Core collapse is an abrupt and irreversible marker for the end of a massive star's lifetime. The core shrinks by a factor of 250 in less than a second, beginning a large shock wave that ultimately destroys the star.

A couple of things happen in extremely short order at this point. The core of the star, which has collapsed down into neutrons, will abruptly stop collapsing as the core density becomes so high that a new force can resist gravity: the strong nuclear force. This force appears because the neutrons are now so tightly packed together that they resemble an enormous nucleus. The strong nuclear force resists the inward force of gravity like an extremely stiff spring, so while the neutrons can be compressed a little past where they would like to be, there's an outward bounce as the strong nuclear force wins against gravity.

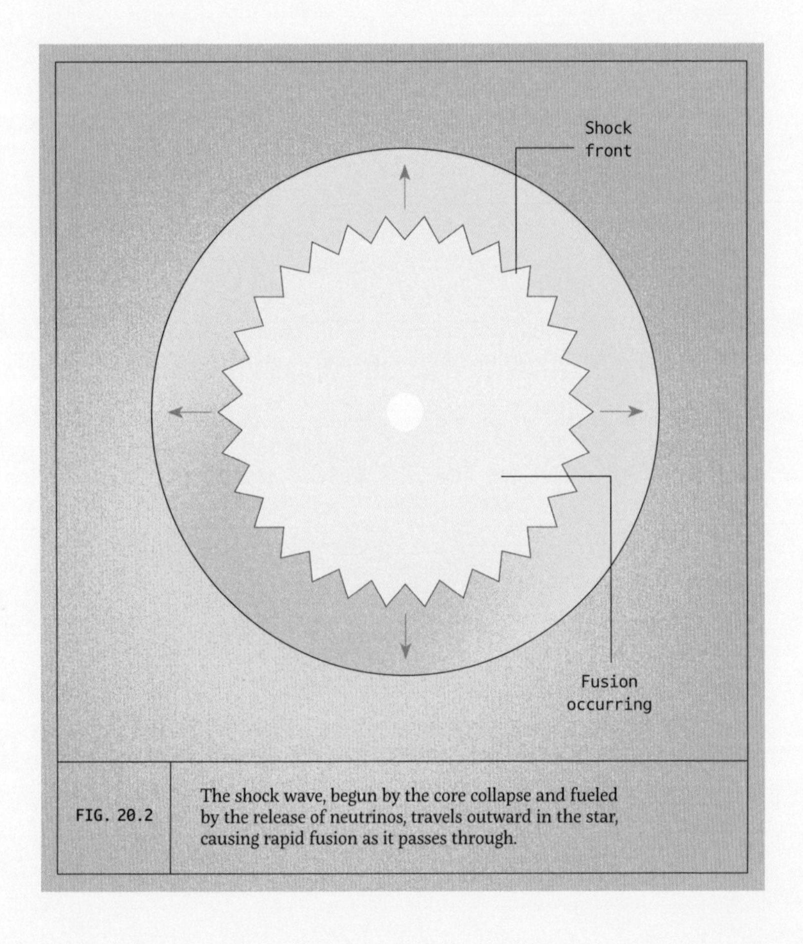

FIG. 20.2 The shock wave, begun by the core collapse and fueled by the release of neutrinos, travels outward in the star, causing rapid fusion as it passes through.

Simultaneously, the outer layers of the star, which are used to being supported from the inside, have just had that internal support removed. The collapse of the core means that the size of the core has reduced from a radius of about 6,200 miles (10,000 kilometers) down to about 25 miles (40 kilometres), and it's happened so fast that the outer layers of the star weren't moving in synchrony. Devoid of any support, the star begins to collapse.

The rebound of the neutrons in the core creates a shock wave going outward, and this collides with the rest of the star falling inward, compressing and shocking even more of the gas (fig. 20.2).

The rebound alone isn't enough to detonate the star, but we haven't yet dealt with the neutrinos that were made during the core collapse. The star has successfully created a very high temperature and pressure shock wave, and while neutrinos usually don't interact with material very much, they do interact with this shock wave. The neutrinos carry so much energy that depositing even a small fraction of this energy into the stellar shock wave is enough to ensure not only that the shock wave makes it to the surface, but that it does so going at 6,200 miles (10,000 kilometers) per second or more. This is fast enough that the shock wave wins against the force of gravity entirely, and the star is destined to be destroyed in a burst of brilliance.

While the supernova is still underway, with the shock wave radiating outward through what were layers of fusion, temperatures and pressures are temporarily so high in the shock front that fusion can take place at a far higher rate than previously possible. Large volumes of oxygen, for example, are produced in these shock fronts—for a 25 solar mass star, some 3 solar masses of oxygen will be created, along with 0.6 solar masses of neon, and 0.05 solar masses of iron. Elements heavier than iron can also be produced in relatively smaller quantities. These elements usually aren't produced in a star, because creating them consumes more energy than is released (which is a terrible system for a stable star), and to create stable elements needs a large amount of available neutrons. The tremendous heat and density of the shock front provides both.

as a neutron star

To more fully understand a supernova, we can look for the leftovers of the explosion. The entirety of the outer layers of a massive star is lost in the supernova explosion, but in many cases there is a stellar remnant left behind—the remains of the collapsed core itself.

For stars which are above 8 solar masses, and roughly less than 25 solar masses, what is left behind is the pile of neutrons that were once radioactive nickel and iron. We call this a neutron star, though the "star" is still a misnomer here. There's no fusion happening, and the neutron star isn't even made of the usual collection of plasma. It's almost exclusively neutrons.

Neutron stars are fantastically dense. While many of them are just over 1 solar mass, they're only about 12 miles (20 kilometers) in radius. They're also incredibly round, with the biggest defects from perfectly spherical in the millimeter to centimeter range (typically far less than a quarter of an inch). We also have good evidence to suggest that they host exceptionally strong magnetic fields, far beyond anything we'd experience on Earth. A typical MRI scanner has strengths of 0.5 to 1.5 Tesla, and at that level we need to be extremely careful about having any kind of magnetic materials in the same room. For a neutron star, the magnetic field strength can be well over 100 million Tesla (fig. 21.1).

Because of their small size, they can be hard to observe directly. Even for nearby supernovae remnants, it often takes a long time to identify the neutron star in the center of the supernova debris. A supernova that exploded in 1997 in the Large Magellanic Cloud, a tiny companion to our Milky Way, only had the neutron star directly imaged in 2024.

The best way to spot a neutron star, if it's not very nearby, is to make use of the strong magnetic field and the fact that neutron stars spin, often extremely rapidly. The spinning neutron star is able to spiral electrons along the magnetic field, and this spiraling generates long wavelength light, detectable with radio telescopes. If these particles spiral out along the magnetic poles, we get a strong collection of radio light beamed our way. The magnetic poles are not always aligned with the spin of the object, and so these beams

FIG. 21.1 A neutron star, the leftovers from a supernova explosion, is an extremely dense, extremely magnetized object, most easily visible if the magnetic poles happen to point our way.

often sweep broad paths across the sky, flickering in cadence with the spin. When they were first discovered by Dr Jocelyn Bell-Burnell, a PhD student at the time, it was not known that these flickering objects were neutron stars, and so they were dubbed pulsars, to indicate the pulses of light that reached us on Earth.

We now know that pulsars are simply a subset of the neutron star population in which the geometry of the system has lined up so that we can see the flash of light as it is pointed our way. The first pulsars discovered sent flashes in our direction every 1.3 seconds, but since that time, much faster pulsars have been detected. The fastest

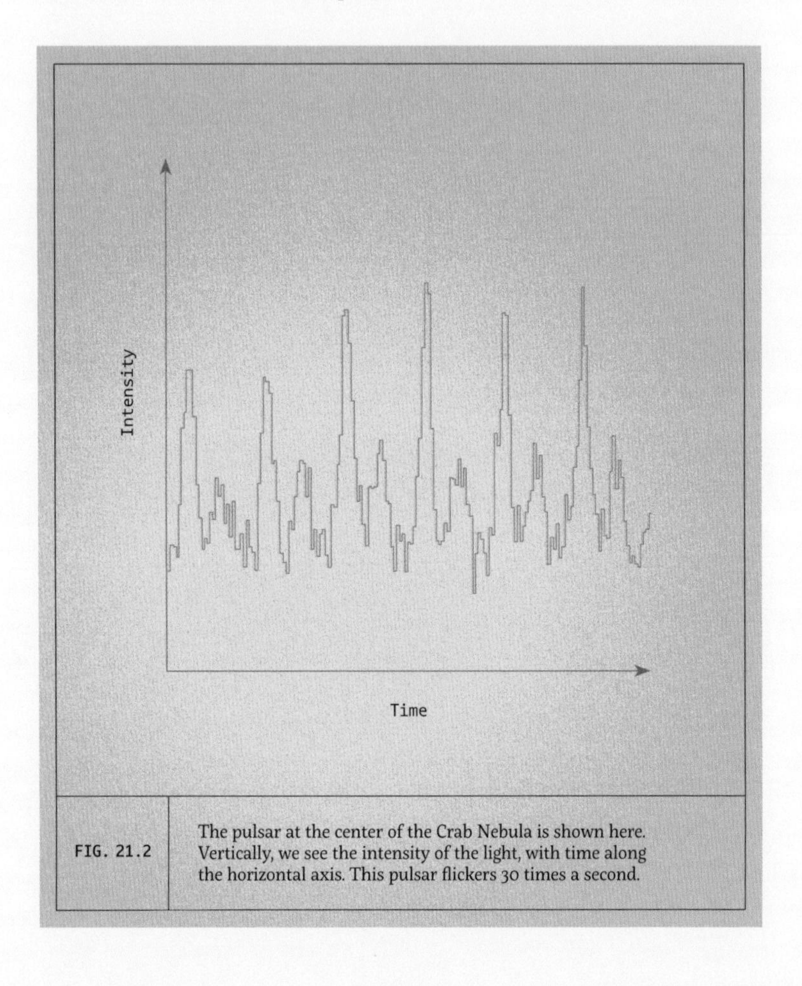

FIG. 21.2 The pulsar at the center of the Crab Nebula is shown here. Vertically, we see the intensity of the light, with time along the horizontal axis. This pulsar flickers 30 times a second.

of the more than 3,400 known pulsars, as of 2024, spins 716 times every second, with more continuously being discovered (fig. 21.2).

One of the most dramatic pulsars we can see is the neutron star at the center of the Crab Nebula. This nebula is the remains of a supernova that exploded in 1054 CE and was observed as a "new star" by most human civilizations at the time. There have been considerable efforts to pin down exactly when this star exploded, and to search early records for observations of the supernova. The neutron star itself can be observed in detail with modern instrumentation, and it currently flickers at a rate of 30 times a second. It is the brightest of the pulsars detectable in visible light, largely because it's quite close to Earth, at only 6,200 light years from home.

Most pulsars serve as extremely stable self-regulating clocks, but a small fraction of them sometimes speed up, in an event known as a glitch. These glitches are, on a human time frame, very small changes to the spin of a neutron star, but because the neutron star is normally so precisely regular, even tiny deviations from their expected cadence are both noticeable and puzzling. These are interesting behaviors to study, because they're likely to be telling us something about the internal workings of the neutron star. One explanation for many glitch events is a kind of "crustquake," where the solidified outer crust of the neutron star undergoes an earthquake-like abrupt shifting in shape, which would then change the rate of spin. Not all pulsar glitches can be explained this way, so it's likely that more than one thing is happening.

PLATE 6

From the history books to modern images

A supernova leaves behind a dramatic nebula, dispersing the former star at high speeds. In this case, a neutron star is all that remains of the star. This particular supernova remnant, the Crab Nebula, was observed by many people in the year 1054 CE, and would have been brighter than everything in the sky except the Sun and the full Moon.

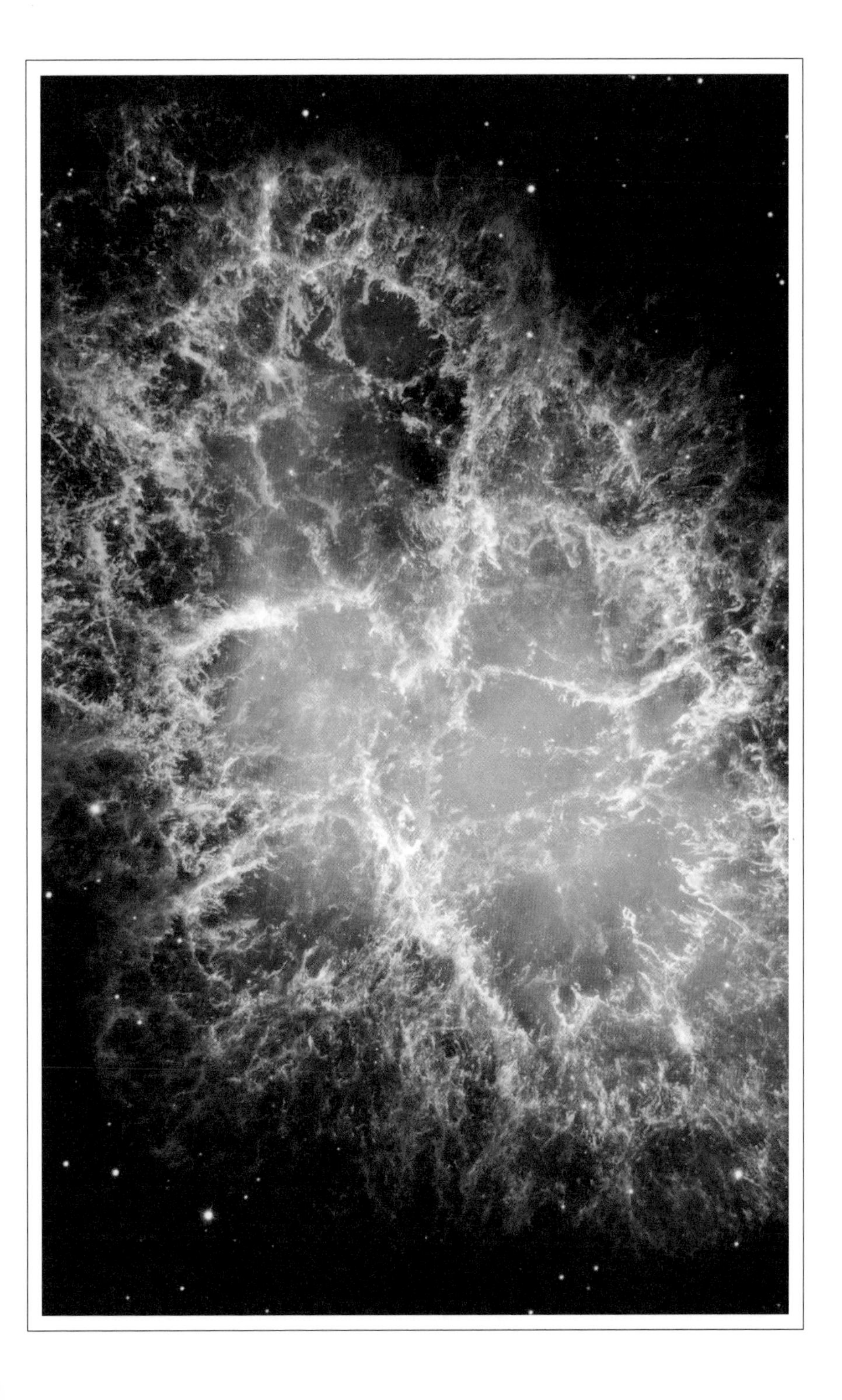

as a black hole

To know a star, we can also look at the only other possible outcome of a supernova explosion: a black hole. Black holes arise when we look at the supernovae ending the lives of the most massive stars. In these systems, the core, prior to shattering, reaches a mass greater than three times that of the Sun. Core collapse undergoes the same process, with a temporary halting of the collapse in a proto-neutron star phase which causes a tremendous rebound.

However, in these objects, the strong nuclear force isn't enough to hold the neutron star up against gravity. And in this case, we wind up breaking a lot of mathematics. With no remaining forces operating on smaller scales than the strong nuclear force, once gravity wins out, it has permanently won. The object continues to collapse, unresisted.

Eventually, we create an object so dense that in order to escape its "surface," you'd have to travel faster than the speed of light, which is impossible. This marks the creation of a brand-new black hole, and the contour in space that marks this point of no return is called the event horizon. There's no physical surface here—it's simply a mathematical marker that tells us where the gravitationally doomed limit is (fig. 22.1).

The physical object undergoing collapse has no reason to stop falling inward on itself, and so we presume that it must continue until it cannot collapse any further. It should reach a singular point occupying no volume, but with several solar masses of material crammed inside. This point of infinite density is called the singularity, and understanding its properties runs into mathematical difficulties, since dividing by zero volume to get infinite density is a challenge.

We can, however, observe the impact of this infinitely dense object on anything that comes near it. One of the cleanest ways to detect a stellar mass black hole is to look again for binary stars, where one star has already ended its life and the other remains on the main sequence or in a red giant phase. A black hole, in this case, can be observed in two ways. First, through the induced motion in the bright star: if we see the bright companion to a black hole orbiting an object we cannot see, but whose mass we can infer, we can identify that there must be a black hole, even if it's fully invisible (fig. 22.2).

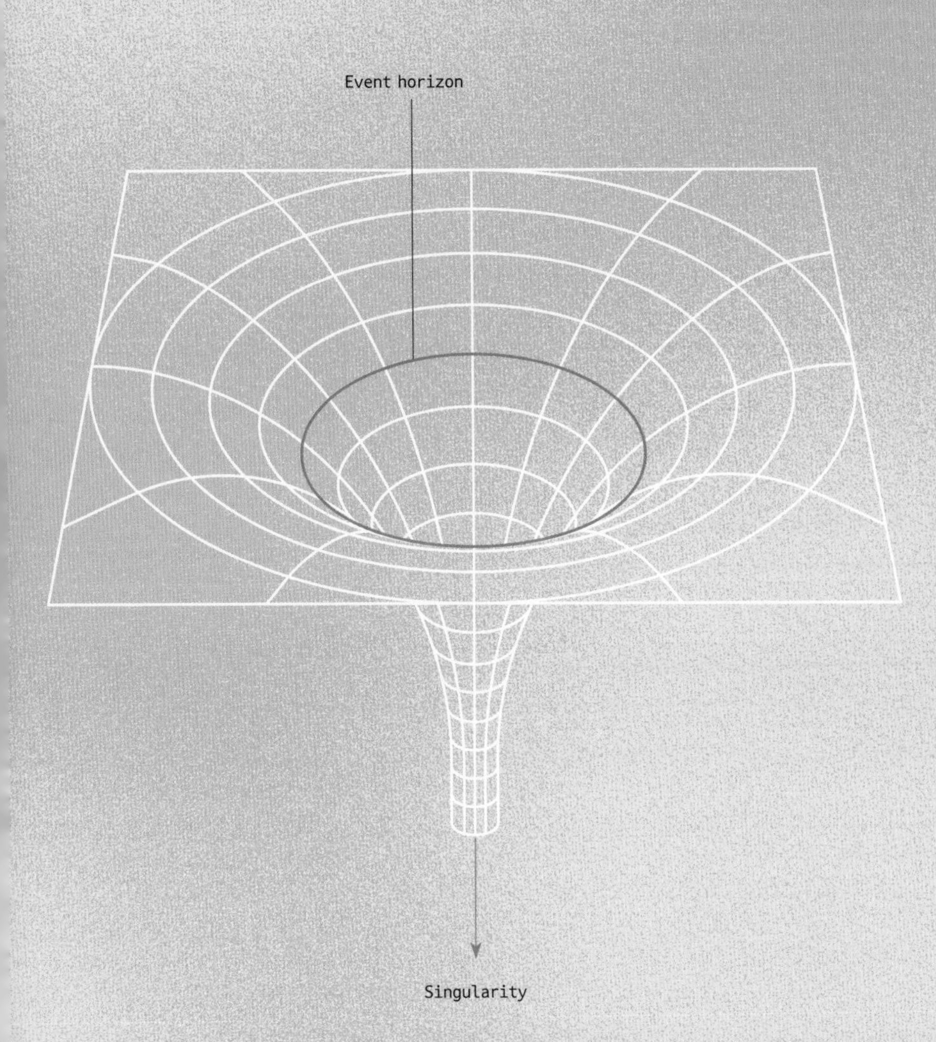

FIG. 22.1	Some of the earliest evidence that black holes were real came from identifying objects that were in binary systems with another star. The black hole often siphons material off of its companion and creates a bright accretion disk of superheated material trying to fall inward.

The second way is easier to see. If our two objects are in a close enough orbit, then much as the white dwarf could siphon off material to fuel a nova event, the black hole can do the same thing, peeling the outer layers of the star away from it, and into a bright accretion disk swirling around the black hole. Black holes aren't terribly efficient at gathering new material past their event horizon, and the material drained off of their companion heats up to a tremendous temperature—up to 180 million Kelvin (100 million degrees Fahrenheit). At these temperatures, the disk glows brightly in the X-ray wavelength range. A lot of the material in this disk never makes it into the black hole at all. A good chunk of it is flung outward in two large jets, perpendicular to the accretion disk. For the lucky fraction of the material that does make it all the way down to the black hole, it will fall inward through the event horizon, and eventually settle into the singularity, adding a tiny amount of mass to the black hole.

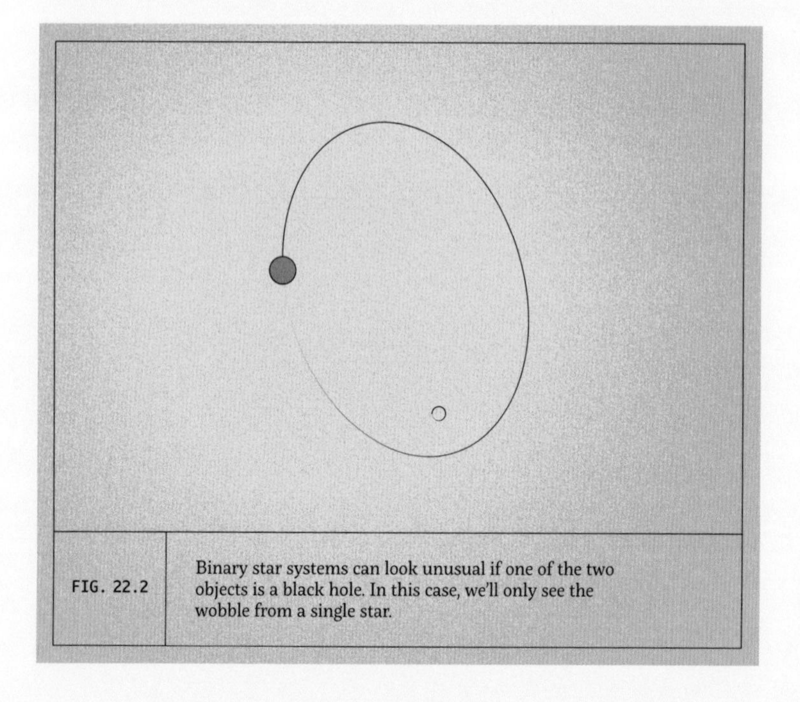

FIG. 22.2 Binary star systems can look unusual if one of the two objects is a black hole. In this case, we'll only see the wobble from a single star.

Because gravity is so extreme when we're close to the event horizon, there are intense gravitational tidal forces. Much in the same way that the Moon pulls on the near side of the Earth more strongly than the further side, because gravity is stronger between any two objects the closer they are, black holes can do the same things to any material close enough. The near side of any object falling toward a black hole will have a much stronger gravitational pull on it than the far side, and this means that the object will stretch out. The near side gets even nearer, which exposes it to an even stronger gravitational force, and so the stretching becomes more extreme. This process is scientifically known as spaghettification.

It's worth noting that you really do need to have material very close to the black hole for this effect to be noticeable. At any large distance, black holes behave exactly the same as any other massive body, which is why they can operate like any other star in a binary system, separated by millions of miles/kilometers. The event horizon of a 10 solar mass black hole is only 19 miles (30 kilometers) in radius, so for spaghettification and other extreme events surrounding a black hole, we'd have to be only tens of miles away from the singularity. Given a wide enough berth, a black hole is merely a fascinating pucker in gravity.

as an unsteady object

Some stars are anything but stable sources of light. While we do expect the vast majority of stars, for the overwhelming majority of their lifetimes as luminous objects, to be stable, there are a lot of stars that don't fit into this category. And the longer we look, the more we find.

One of our best current methods of finding variable stars in the sky is the Gaia satellite, whose mission goal is to accurately map the locations of a billion stars in our own galaxy. Since this takes time to complete, it has compiled a huge catalog of stars whose brightnesses have changed over the course of the years it's been observing the sky. The most recent data set, released in 2022, contained 1.8 billion objects, of which some 10 million are likely to be galaxies outside our own, and another 10.5 million are variable sources.

Variable stars are classified based on their behavior. How quick is the cadence of their variability? Does the variability repeat, or is it a one off? Is the brightening and dimming following a regular pattern, or is it a little more random? Does the star gently brighten and fade, or does it sharply jump in brightness and then fade over time?

Technically, any object that changes its intrinsic brightness, even once, could be considered a variable star. This means that events like novae and supernovae are considered a kind of variable star, even though in the case of a supernova, it will definitely only happen once. These events are deemed cataclysmic variables, though only the repeating novae will have a repeated cadence to their explosive variability (fig. 23.1).

We also have a class of star called a flare star. These are usually low mass stars, on the main sequence, which periodically have tremendous solar flares. The outburst of light from the flare is sufficient to brighten the star, which then subsides. These flare stars do not usually have a predictable pattern to their brightenings, because the process underpinning the flare is a magnetic one, and this is fairly erratic. The nearest star to our Sun, Proxima Centauri, has been observed to flare in this way, sometimes producing flares

FIG. 23.1	Low mass stars seem to be more active than our Sun, producing relatively powerful coronal mass ejections and other eruptions from their surfaces, changing their brightness as seen from Earth.

as energetic as the Sun's, in spite of being only about a tenth of the Sun's mass. These flares—which produce a lot of X-rays—are likely to destroy any kind of atmosphere that might have formed around a planet. Once the atmosphere is gone, the UV radiation released by the flares ought to sterilize the surface of the planet—not ideal if you're looking to grow things.

Most of the rest of the stars whose light output changes with time are physically changing in size, and are definitely no longer on the main sequence of star formation. These are usually stars in some stage of their red giant phase. One of the more

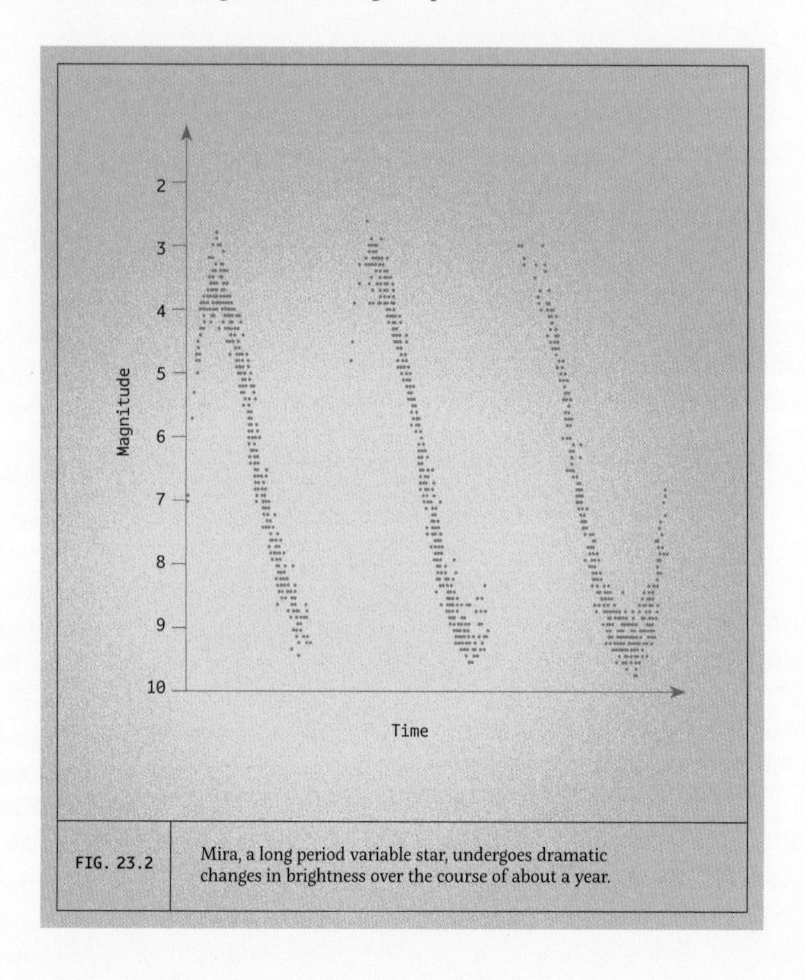

FIG. 23.2 Mira, a long period variable star, undergoes dramatic changes in brightness over the course of about a year.

dramatic examples of this kind of star, aside from Betelgeuse, is Mira, in the constellation of Cetus. This is a relatively slow variable star, cycling through bright and faint intensity once every 330 days. It lies on the asymptotic giant branch, so it is undergoing fusion in shells around the core, and we expect it to turn eventually into a white dwarf. Mira is the prototype of the "Mira variable" class of stars, which are particularly striking in their variability. Mira itself varies by six magnitudes in brightness, meaning that it pulses in and out of naked-eye visibility in the night sky (fig. 23.2).

At the other end of the flickering cadence are RR Lyrae stars, which usually vary on timescales of less than twelve hours. Oscillating in brightness by about half a magnitude, they're not quite as dramatic as the multiple magnitudes of Mira variables. RR Lyrae stars, while having exited the main sequence, are usually quite low in mass, meaning that they are quite old. In general, they tend to be 60–80 percent the mass of the Sun, which makes them about 10 billion years old.

Of course, there are exceptions to every classification scheme, and luminous blue variables are one of these. The most dramatic example seems to be Eta Carinae, which underwent some kind of massive eruption in the late 1830s, forming the Homunculus Nebula around itself, brightening temporarily by around four magnitudes, and then subsiding. This massive outburst probably resulted in the loss of 3 solar masses-worth of material from the star. Extremely massive, blue, and luminous stars like Eta Carinae will probably undergo a supernova explosion in the future.

as an eclipsing binary

A closer look at the Demon Star, more formally known as Algol, reveals a triple-star system. The two most massive stars orbit very close to each other, and the third is in a far more distinct orbit. Algol is also a very good example of how we can observe a star that is variable in brightness.

The two massive stars pass one in front of the other, from our perspective. This is called an eclipsing binary system, and in these instances the eclipse will be regular and dramatic. Instead of the rise and fall of an intrinsically variable star, an eclipsing binary starts with a completely constant amount of light, which is the sum of the two stars. The eclipse is abrupt, with one star partially or fully blocking the light from the other. In extreme cases, the amount of light we receive from them will stabilize at some lower level, and then as the background star re-emerges into our line of sight, the light will return to its original level. The most common way of representing these systems is through a light curve, which is the fraction of the maximum amount of light we receive from the system, as a function of time (fig. 24.1).

A lot of information can be learned about the stars from examining their eclipses, and so while they're geometrically uncommon, they're fantastically helpful. And since we've done so much hunting for interesting stars, we've also found a lot of them—there are at least 500,000 in the Gaia data so far; the planet-hunting satellite Kepler found 2,878 of them; and Kepler's successor TESS identified 4,580 more. Neither TESS nor Kepler was actually looking for eclipsing binary stars. These missions were searching for the much fainter signals of transiting planets, but looking for a small planetary flicker will accidentally find a lot of eclipsing binary systems as well.

Since the eclipse is a feature of the orbit of the two stars, it will repeat exactly, and this repetition tells us how long it takes the stars to orbit. If the stars are of equal mass and radius (and therefore temperature), then we should expect the two eclipse light curves to be equal in depth and width. However, in most cases, the two stars are not the same temperature, and so the light curves are not the

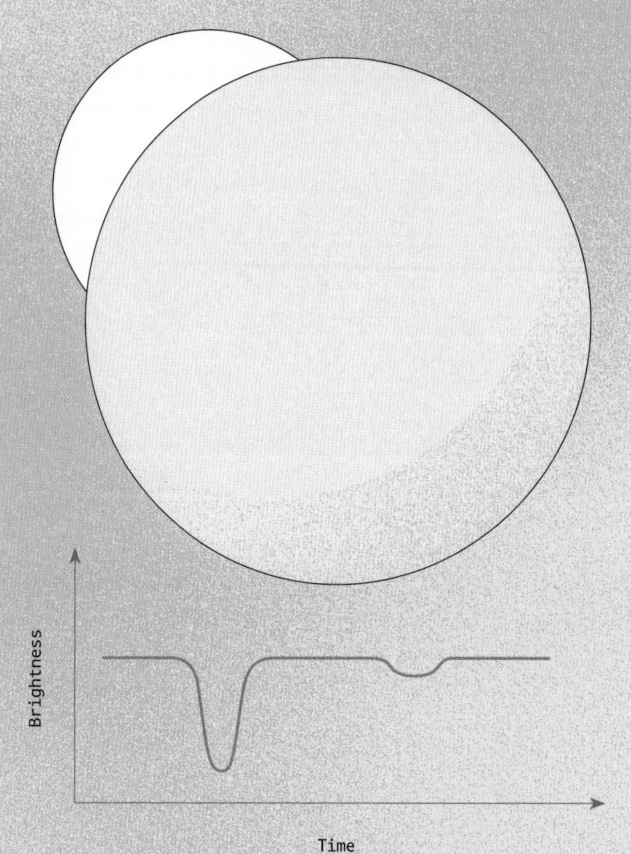

FIG. 24.1	Eclipsing binary stars are those star systems where one star passes directly in front of the other, blocking the light from the background star. This kind of system requires geometrical luck, but is very informative.

same. In these cases, the less massive star (usually smaller, cooler, and therefore dimmer) will block some fraction of the hotter and brighter star, causing a substantial dip in the total amount of light we receive. When the fainter star passes behind the brighter one, however, while the total amount of light drops, it's less substantial a drop, since it's a less luminous object being obscured.

With the velocity of the two stars, we can use the length of time it takes for the eclipse to reach its maximum value, or to come back up from the deepest point of the light curve, to find the radius of each of the stars directly (fig. 24.2).

Algol, as a result, we know quite a bit about. The hotter, brighter star in the system is about 3.4 times the mass of the Sun, with a smaller companion about 75 percent of the mass of the Sun, orbiting once every 2.8 days in a perfectly circular orbit. The hotter star has a temperature of about 12,500 Kelvin (22,000°F), with the

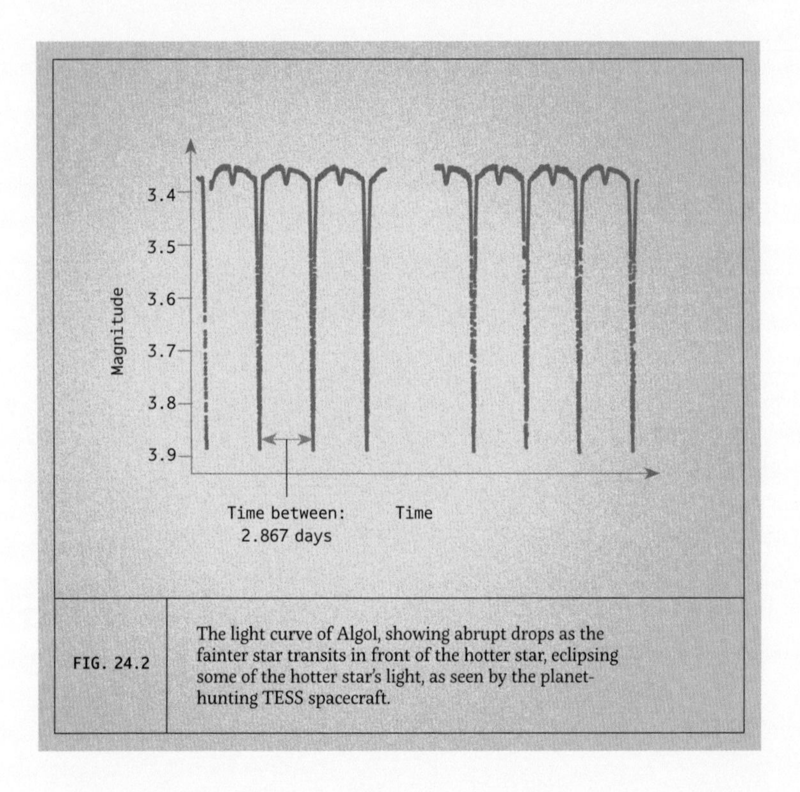

FIG. 24.2 The light curve of Algol, showing abrupt drops as the fainter star transits in front of the hotter star, eclipsing some of the hotter star's light, as seen by the planet-hunting TESS spacecraft.

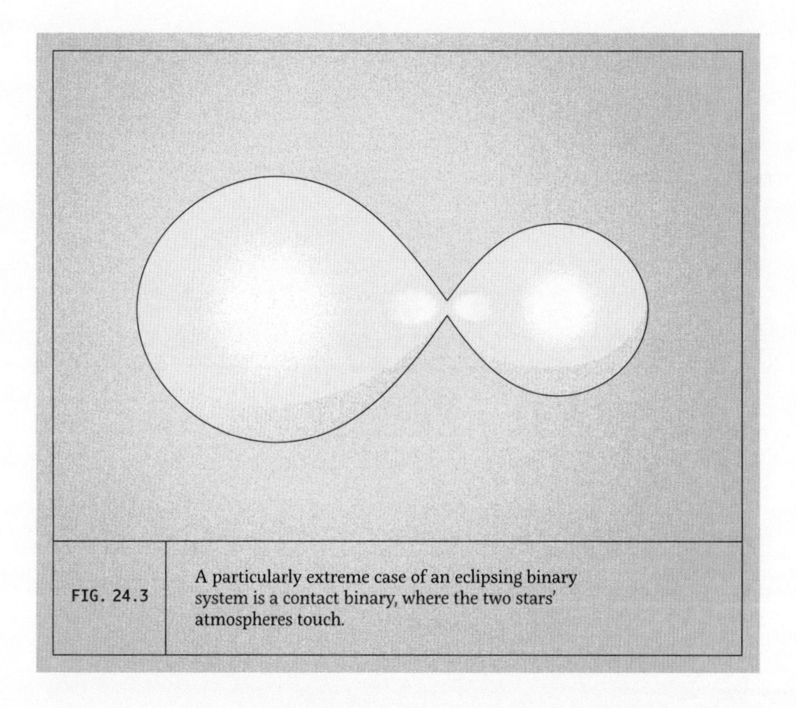

FIG. 24.3 A particularly extreme case of an eclipsing binary system is a contact binary, where the two stars' atmospheres touch.

fainter companion considerably cooler at 4,900 Kelvin (8,360°F). Further research has shown that the two stars are close enough together that they have undoubtedly passed material between them. Radio observations show a stream of gas, passed along magnetic field lines, from the low mass star to the higher mass star.

This passing of material between the two stars is expected, considering how close they are to each other, but it can get more extreme than this. There's a class of eclipsing binaries called contact binaries. These stars are so close together that their atmospheres actually touch (fig. 24.3). Rather than a double-lobed blob of a star, usually these stars are observed as a very tight eclipsing binary system, where their orbits are so small, and their radii sufficiently large, that the surfaces of the two stars must be in contact with each other. These kinds of stars may eventually merge together into a single star, but if the atmospheric contact is minimal enough, this outcome is not guaranteed.

PLATE 7

An explosive aftermath

Eta Carinae, a luminous blue variable star, underwent a dramatic eruption event in the late 1830s, creating the Homunculus Nebula surrounding the star itself.

as part of the Milky Way

Looking at the patterns in our night sky teaches us what structures the stars can build. All of the stars visible by eye are part of a much grander structure: our own Milky Way galaxy.

Early surveys of the skies revealed that the stars are not equally numerous in every direction, which is our first hint about the structure of the Milky Way. In dark skies, this is apparent even to the unassisted eye, with a band of what may appear to be high, thin clouds crossing the sky. Instead of clouds in our own atmosphere, what we're actually seeing is a crowd of distant stars, all contributing a little bit of light in our direction.

The Milky Way existing as a band in our skies is a clue to its structure, and our position in it. If the Milky Way were a spherical object, we would not expect it to present a thin edge to us. Therefore it cannot be particularly spherical. If we were in the center of such a sphere, we would see roughly equal numbers of stars everywhere we look. If we existed at the edges of a spherical galaxy, we would expect to see an extremely dense sky at some parts of the year and relatively fewer stars at other times, when our night sky points us away from the densest parts of the galaxy (fig. 25.1).

Early attempts to find the shape of the galaxy relied on this principle, but were hampered by the Milky Way not being a pure collection of luminous objects. In between the stars there is gas and very fine dust, and these block the light from the stars behind the dust clouds. Nonetheless, judging from the narrow band where the stars are densest, we must be in some kind of flattened object.

The Milky Way has been a uniquely challenging object to map out, purely because we're embedded within it. For objects where we can see the whole thing at a glance, figuring out their structure is straightforward: simply look at it. But for us, nestled within the Milky Way, identifying its actual structure has been anything but easy, and we are still finding new features as we develop new instrumentation.

We have learned that our Milky Way is shaped like an extraordinarily flat disk, with dense "arms" of stars in higher densities lying within that disk, spiraling out from the center

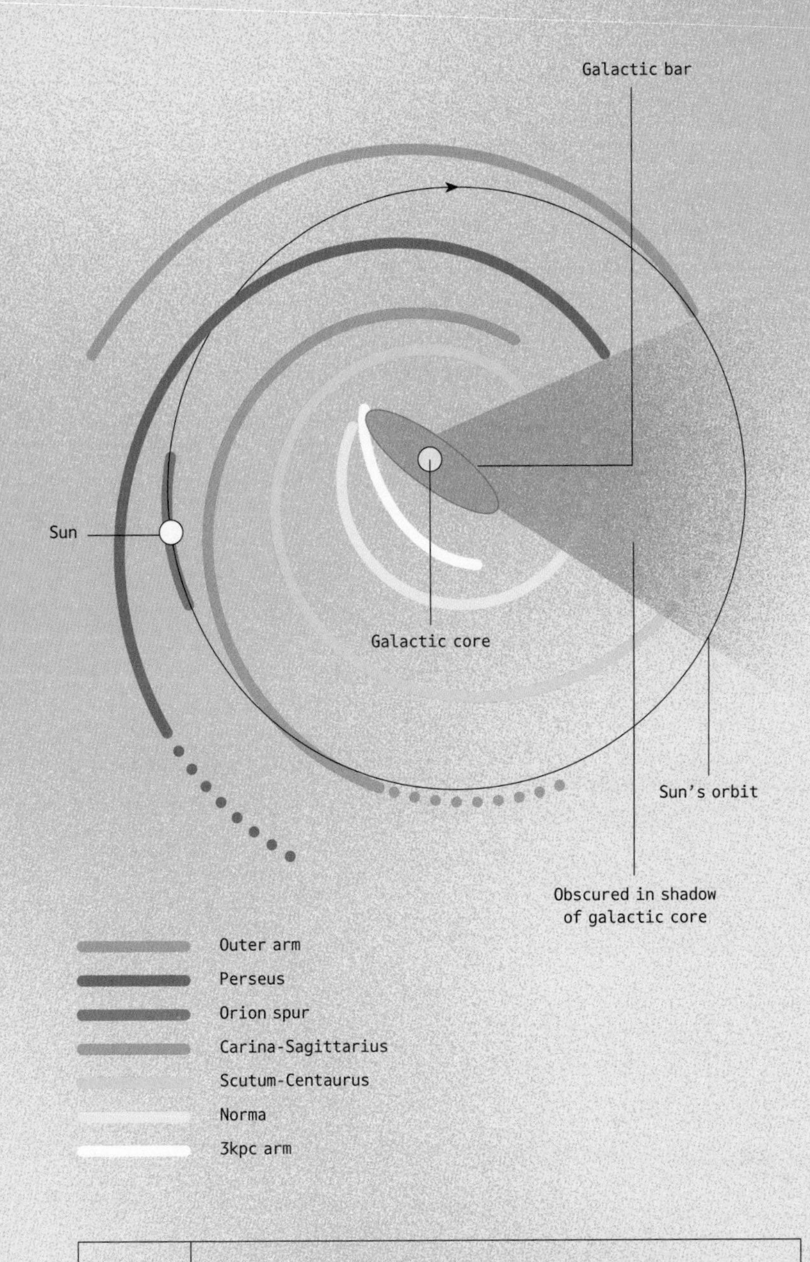

FIG. 25.1	Our Milky Way, as we currently understand it, has a large nuclear bar across its center, and spiral arms extending outward once the bar ends. Our Sun is moderately removed from the center of the galaxy, in between two more major spiral arms.

of the galaxy. Our own Sun lies some 26,500 light years from the center of the galaxy, in an offshoot of one of the major spiral arms. It wasn't until 1991 that we had solid evidence for the bar in the center of the Milky Way, a structure that looks out of place in an otherwise spiraling round galaxy. It's a roughly straight line, extending some 16,000 light years out from the center of the galaxy, and at a 30-degree angle to our vantage point from Earth. This places the Milky Way among a class of galaxy known as a barred spiral galaxy (fig. 25.2).

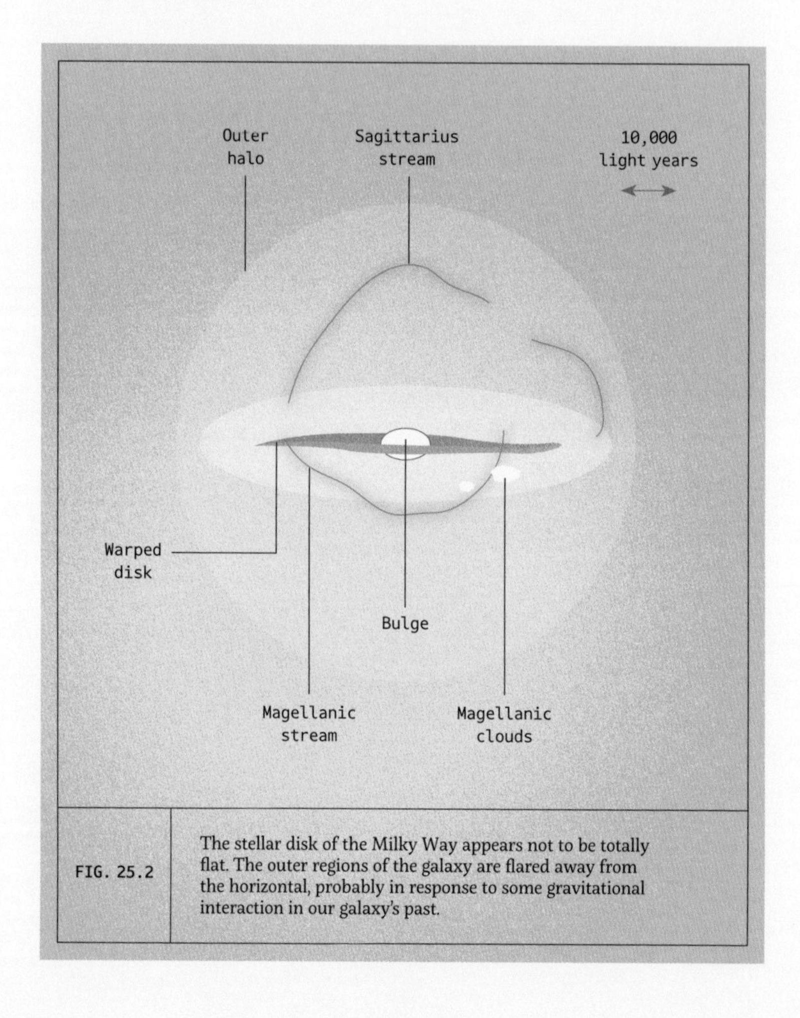

FIG. 25.2 The stellar disk of the Milky Way appears not to be totally flat. The outer regions of the galaxy are flared away from the horizontal, probably in response to some gravitational interaction in our galaxy's past.

As we examine the Milky Way's stars more carefully, we've been able to notice some other unusual features, traced entirely by the faint signature of distant stars. The first is that while the stars in the Milky Way's disk are thinly collected, that disk is not perfectly flat. It is warped, with the disk flaring up on one side and down on the other, and more dramatically the further away from the center we get. Current theories for this warping most commonly point to some kind of gravitational interaction with the Milky Way—either an ancient collision, or a collision with one of our small satellite galaxies. In either case, the object could have disturbed what would have previously been a flat disk, and set up a gravitational wobble. In the case of a small satellite galaxy being the culprit, it may have passed through the outer disk of the Milky Way, setting up a distortion we can still see in the outer galaxy's stars and how they move. One model indicates that the best fit for this warp is a recent, small interaction, implying that the warp may be a relatively new addition to our galaxy's list of features.

Close examination of the region around our galaxy has provided even more evidence for small interactions between the Milky Way and smaller companions. A hunt for stars where we weren't expecting to find them has led us to the Sagittarius Stream. The stream itself is made up of stars that once belonged to the Sagittarius dwarf galaxy, but whose interactions with the Milky Way have disrupted it enough that it has lost stars along its long, looping path high above the disk of our galaxy.

PLATE 8

A view of our home

The Milky Way, as seen by the
European Space Agency's Gaia spacecraft,
shows the density of stars in the sky,
tracing nearly 1.8 billion individual stars.
Dark dust lanes obscure the bright
disk of our galaxy.

by its petal-like orbit

We can also watch how the stars move within our galaxy. As with many other detailed properties of galaxies, this is most easily done within our own galaxy, and in particular, with our own Sun and nearby stars.

To know the orbit of a star, we need to be able to pinpoint both its position and its velocity—both the speed and direction of its motion. All objects under the influence of gravity alone should orbit in some kind of ellipse, with a circle being the simplest possible solution. Circular orbits or near circular orbits are common, but not guaranteed, and many stars orbit the centers of their galaxies in longer, more oval-shaped orbits.

Many stars orbit in very slightly non-circular orbits. They're not strictly a circle, but they might resemble one if you squint. Our Sun is on one of these orbits, and we have measured our solar system's velocity relative to the center of the galaxy to be about 450,000 miles per hour (724,000 km/h), taking about 230 million years to complete one loop around the center of the Milky Way (fig. 26.1).

We have some fun language to describe these slightly non-circular orbits. Perigalacticon is the point in the orbit of a star where it comes closest to the galactic center. Apogalacticon is the opposite: the most distant location from the galactic center. A typical star in the outer regions of the galaxy will usually not vary terribly far in its distance from the center. For the Sun, this variation is thought to be only a few percent, which means our orbit is very close to circular. But drift of a few percent of 26,000 light years can still be a substantial change in the distance from the galactic center in terms of light years.

There are two more ways for these orbits to become more complicated. The first is that when we think of a circular orbit, we usually think of a flat, two-dimensional orbit. But there's nothing about orbiting in outer space that says the orbit actually has to be perfectly flat with no vertical motion (relative to the disk of the galaxy). In fact, if there is a little bit of upward or downward motion, we'd expect the orbit to begin to bob up and down, oscillating vertically through the disk of the galaxy.

FIG. 26.1	Apsidal precession is the result of a not fully isolated system, and rotates an elongated orbit over time, drawing a gradual rosette out of the orbital path.

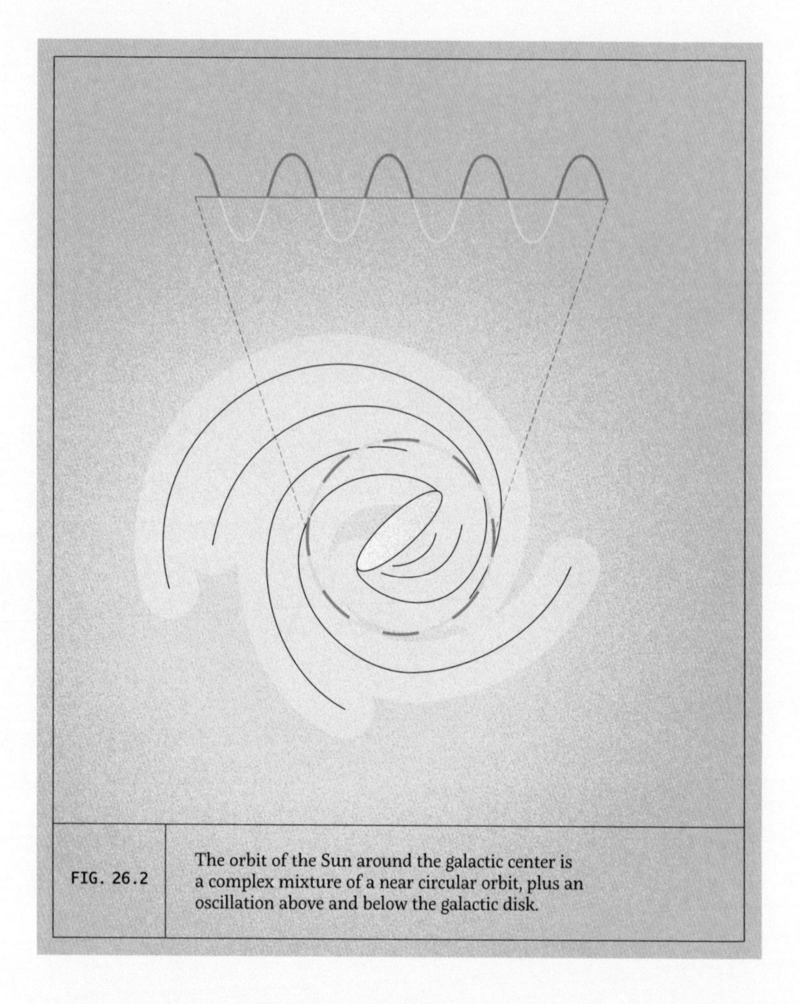

FIG. 26.2 The orbit of the Sun around the galactic center is a complex mixture of a near circular orbit, plus an oscillation above and below the galactic disk.

As the star finds itself above the galactic plane, gravity will tug on it back "down" to the plane. With nothing resistive to slow this motion, the star will overshoot and wind up below the disk of the galaxy. Gravity will now pull the star back "up," and this process will repeat. The vertical oscillation of our Sun's orbit around the Milky Way is substantially faster than the time it takes to complete an orbit. It completes one vertical bounce every 26 million years, so it'll undergo about 8.5 vertical bounces for every orbit around the galaxy's core (fig. 26.2).

The second way that a star's orbit can change with time will be more obvious the more elongated the orbit is. It's a change in the angle of the long direction of the orbit, relative to some fixed perspective. For nearly circular orbits, this is hard to see, but for elongated orbits, a shift in this angle means that the star will gradually trace out an elaborate rosette shape. The shift is not usually very strong, but over time and many orbits, it can cause a rearrangement of the orbit relative to whatever arrangement you might like to set. This change is called apsidal precession, and it's the same kind of shifting we see when a spinning top begins to fall, and the direction of the top begins to swirl around in its own circle. The planets, and indeed the Moon around the Earth, also undergo this kind of precession. It occurs whenever two objects orbiting each other—in this case, the star and the center of the galaxy—are not fully alone, gravitationally. The orbit of the star is perturbed by the presence of the other stars in the galaxy, and so it gets nudged enough to shift the orbit, very slowly.

As we venture further into the center of the Milky Way, the orbits of the stars get less and less orderly. We switch from being largely dominated by a very thin disk of stars into a zone that has stars in a different structure—the bulge. In the bulge, stars generally have more random orbits. While they continue to orbit around the center of the galaxy, they're much less aligned with each other, and orbits tend to carry each star vertically away from the disk in a much more dramatic fashion than the bobbing of our own Sun. With these more random orbits, precession can be much more noticeable—though still long on a human timescale.

as part of a galaxy

We can also look outward at the other kinds of galaxies that stars illuminate for us. Much as our home galaxy is traced by the light from the stars, so is every other galaxy. With other galaxies we are often able to see more of their shape all at once, because we have an external perspective on them. It was realized in 1923 that the Milky Way was not the only galaxy in the Universe, and no time was wasted in beginning to classify the galaxies. In 1926, Edwin Hubble published a classification system, setting out what he saw as the broad classifications of galaxy types.

We still use a modified version of these types today. The major dividing line is between spiral galaxies, like our Milky Way, and elliptical galaxies, which are much redder and rounder systems. Elliptical galaxies tend to be much more massive than spiral galaxies (on average), and the lack of blue stars tells us that their stars formed long ago. Any blue stars that would have formed inside that galaxy alongside the lower mass redder stars will have long since ended their lives. Hubble subdivided the elliptical galaxies by their apparent elongation, using the relative lengths of their short side compared to their long edge. Unfortunately, shape alone may not tell us very much. An elongated, rugby-ball-shaped galaxy may look round if you happen to see it nose-on. You'll need more information to know if the galaxy is round by viewing angle only, or truly spherical (fig. 27.1).

For the spiral galaxies, there were two more classification criteria. The first is whether the galaxy is barred, like the Milky Way, or unbarred, like our nearest neighbor Andromeda. Unbarred spiral galaxies have spiral arms that extend all the way inward to the nucleus of the galaxy, without the interruption of the bar structure.

Within the barred and unbarred galaxies is another classification based on how tightly wound the spiral arms seem to be. However, the tightness of the spiral arm is a somewhat subjective measure of a galaxy, which Hubble himself acknowledged as being "rather arbitrary," and so this has fallen out of favor. We now use more quantitative metrics. Spirals, in general, tend to be bluer than their elliptical companions, and usually less

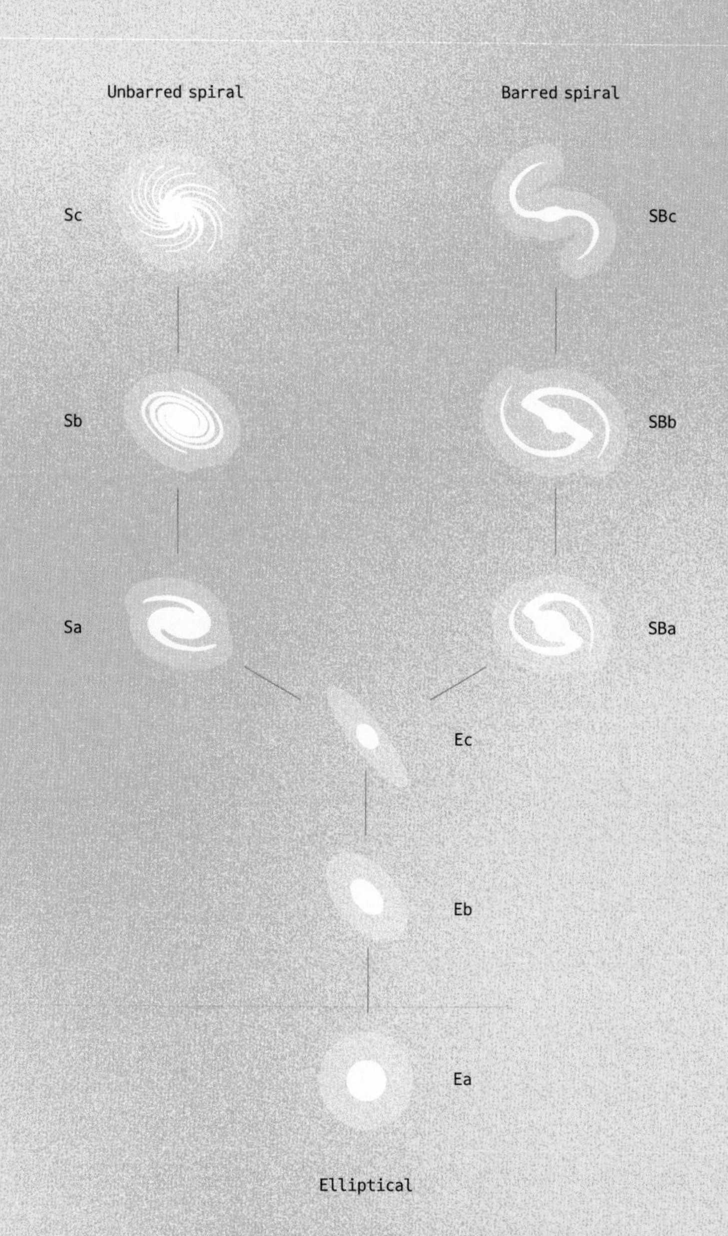

FIG. 27.1	Hubble's classification scheme divides galaxies into ellipticals and spirals, and spirals further into barred and unbarred galaxies. Ellipticals are subdivided according to their elongation, and spirals on the tightness of their spiral arms.

massive. They are also much more sensitive to the angle with which we view them. Seeing a spiral face-on gives a direct view into the spiral arm structure of the galaxy. Viewing it from exactly edge-on gives us a sense of what the dust inside that galaxy might be doing, and also how thin these galaxies can be. More usually, though, we see the spiral galaxy at some kind of angle, as indeed we see Andromeda, avoiding both extremes in perspective (fig. 27.2). The last classification Hubble created was that of "irregular"

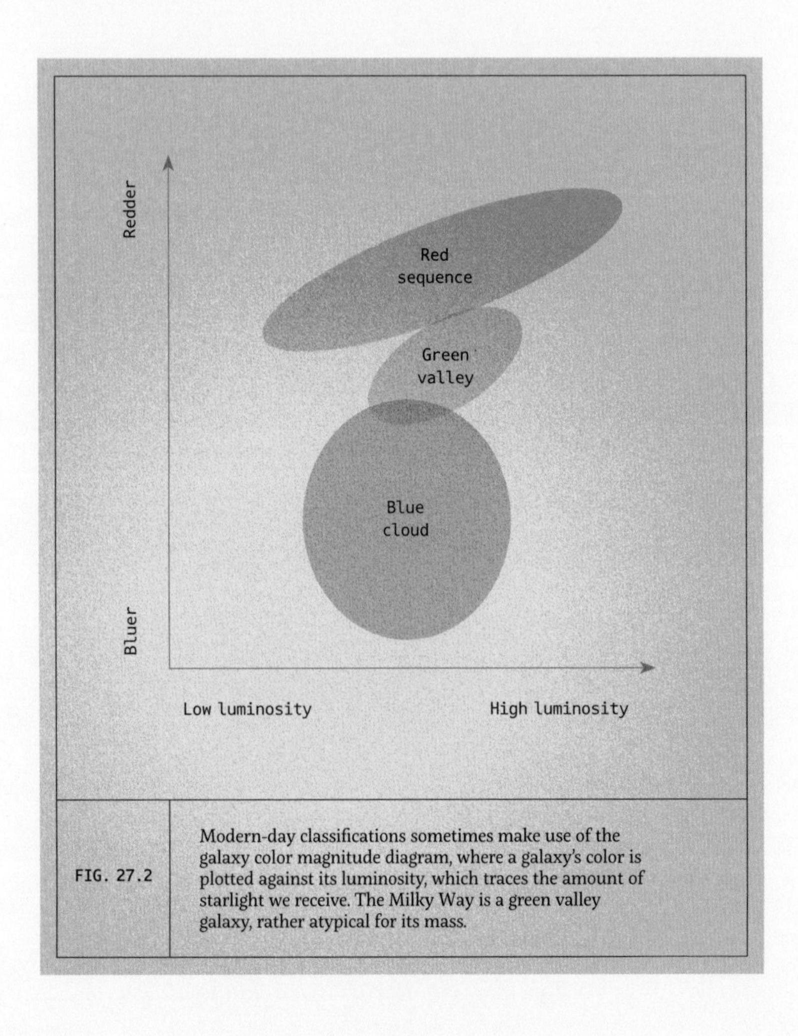

FIG. 27.2 Modern-day classifications sometimes make use of the galaxy color magnitude diagram, where a galaxy's color is plotted against its luminosity, which traces the amount of starlight we receive. The Milky Way is a green valley galaxy, rather atypical for its mass.

galaxies which had no particular symmetry, and this was basically a catch-all term for all the galaxies that didn't fit cleanly into either spiral or elliptical categories. These too come in multiple forms. Some are extremely low mass objects, forming a large number of stars, and don't seem to be massive enough to create the kind of ordered rotation of a spiral galaxy. There are other galaxies, however, that have plenty of mass, but still somehow are extremely chaotic in shape. We now believe these massive irregular galaxies are undergoing tremendous collisions, which will shape them for the rest of their existence. Massive irregular galaxies are often forming a huge number of stars, with a lot of dust obscuring the details of what may be happening underneath. But with examinations of enough galaxies, and with enough telescopes, we can work out that most of these galaxies probably began their lives as spirals, and had the misfortune to pass too closely to another galaxy. This gravitationally doomed them to fall together over the course of (usually) several billion years.

Even within these classifications, there's a lot of diversity. Some spiral galaxies have strong lanes of dust crossing their bright disks. Some ellipticals show a tiny bit of star formation, or display shells of stars, the remnants of previous collisions with other galaxies, flinging stars into distant orbits. There are galaxies with a thin disk, but without strong spiral arm features. There are spiral galaxies with spiral arms, but with a very red color, indicating that they must have finished forming stars in some non-disruptive way.

within a cluster

Looking at the stars that surround the stars teaches us stellar histories, of how and when they formed. Many stars are formed along with other stars, from a cloud of gas and dust that is too massive for just a single star to be created when it collapses. While a large number of these stellar associations disband relatively rapidly after they form, some of them are able to persist for a long time afterward. There are two major forms of clusters we've been able to see: open clusters and globular clusters.

Open clusters tend to be bright, vivid, and blue collections of a few tens to a few thousand stars. The Pleiades, one of the easiest to observe, is a loose gravitational collection of more than 1,000 stars, the brightest of which are a set of massive stars that we see as the Seven Sisters. We know the cluster must have formed relatively recently simply from the presence of these massive stars, which do not have long lifetimes.

It is thought that open clusters like the Pleiades are formed in a denser configuration than we currently see. They may have formed in an environment that resembles the present-day Orion Nebula, which also has a large number of bright, young stars in its core, illuminating the surrounding environment. Over time, as the stellar wind from the stars continues to fill the area, gas which could have been used to form more stars is swept out and away, removing up to half of the mass in the region. This drop in mass results in the stars being less tightly bound to each other, and so they drift apart. If the stars are close enough to begin with, which they may be in the Orion Nebula, then some of them can remain associated, which is our open cluster (fig. 28.1).

Open clusters are not expected to last forever. Because they are only loosely held together, they are easily disrupted, either from interactions between the stars themselves, or by their orbits through and around the galactic disk. More than 90 percent of open clusters are less than 1 billion years old, and the fact that we see so few older open clusters, perhaps missing their massive stars, indicates that these systems must be disbanding on billion-year timescales. Those which survive tend to be the unusually massive

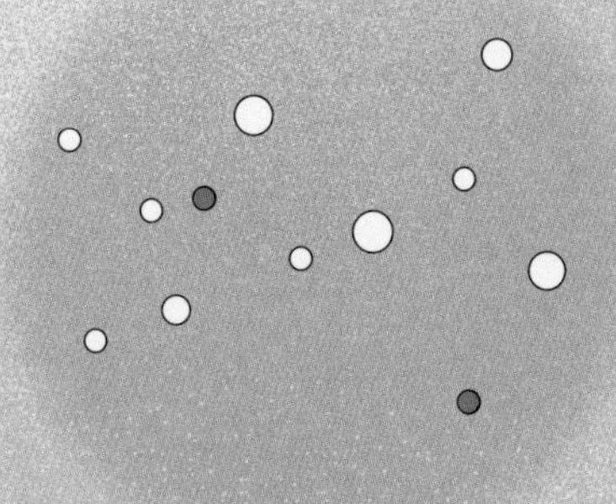

Open cluster

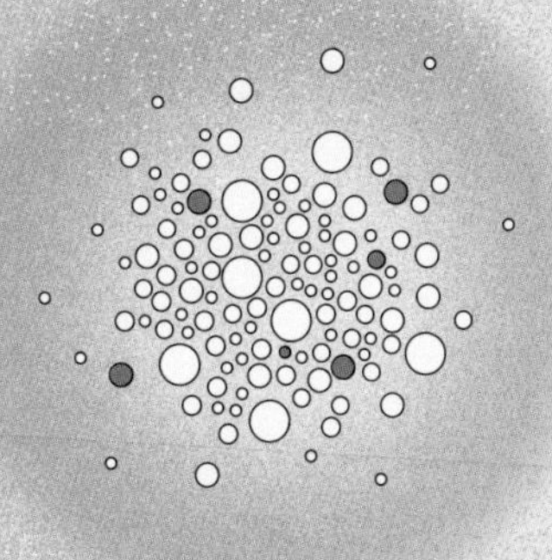

Globular cluster

FIG. 28.1	Open clusters are loose collections of young stars, not expected to last more than 1 billion years in this association. Globular clusters, on the other hand, are ancient, dense, and contain hundreds of thousands of stars.

systems, and even though many of their stars have departed to wander the galaxy on their own, there are enough stars remaining to be visible as a cluster.

Globular clusters, by contrast, are some of the oldest objects in the Universe. While open clusters are usually found near the disk of a star-forming galaxy, globular clusters exist in a more spherical random pattern surrounding the galaxy. Globular clusters are usually far more massive, sometimes containing millions of stars,

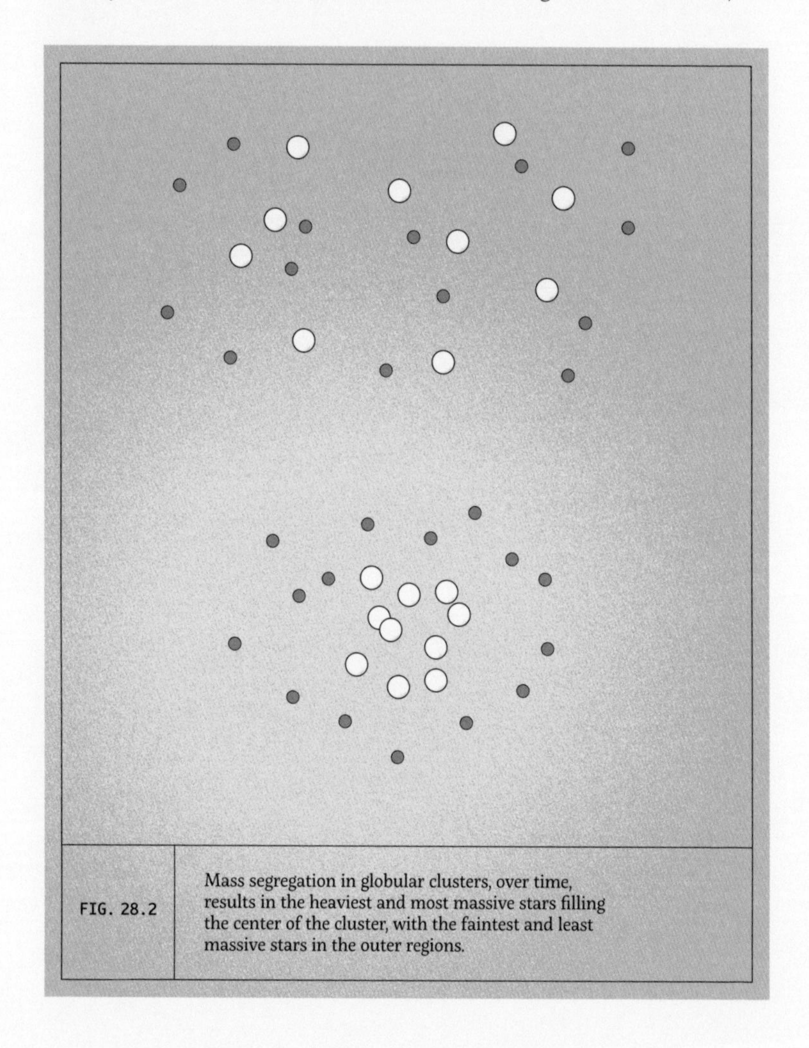

FIG. 28.2 Mass segregation in globular clusters, over time, results in the heaviest and most massive stars filling the center of the cluster, with the faintest and least massive stars in the outer regions.

but distinctly lacking the young blue stars that often dominate the light from an open cluster. Globular clusters, as their name implies, are usually round, and inspections of the orbits of the stars within them reveals that they are truly spherical assortments of stars.

Although we understand the creation of an open cluster, we do not have a good model for how to form a globular cluster. These structures are still mysterious in many ways. We have nonetheless learned that they are more complex than an open cluster in more ways than just their number of stars. Globular clusters also seem to be made up of stars that formed at different times, so rather than representing a single burst of star formation out of a cloud of gas, the globular cluster must have been able to host more than one round of stars forming out of clouds of gas and dust. In any case, globular clusters as we see them now are usually completely empty of gas. Whatever was there has either been used up in forming stars or removed by the stellar wind of the stars that did form.

They are also much more organized than open clusters. Stars within a globular cluster tend to follow a pattern known as mass segregation, where the most massive stars have sunk to the center of the cluster, and the least massive have drifted outward. This organization is rarely seen in open clusters. Mass segregation takes time, and open clusters will typically disperse before it can take place (fig. 28.2).

Because globular clusters are so dense, and filled with so many stars, the upper limit between a globular cluster and what could be considered a very small galaxy is also quite fuzzy. As our understanding evolves, we may be able to create a more meaningful distinction in the future, or perhaps one will be seen more clearly as an extension of the other.

in the structures of galaxies

By looking at how the stars vary in different parts of a galaxy, we can also learn about the structure of the galaxy itself. Differences in stellar age, color, and orbit are part of how we subdivide a galaxy into different components. We have identified a set of components that it seems most galaxies are picked from.

The first and most dramatic structure is that of the disk of a spiral galaxy, such as our own Milky Way. The stars in the disk are a mixture of young and old, with the redder, older stars forming the backdrop for the brighter, bluer stars that formed in more recent times. The vast majority of the stars in our galaxy are in a very thin plane called the thin disk. This thin disk is where almost all new star formation occurs.

Surrounding the thin disk is a less populous but still notable thick disk, where the stars are much older than the thin disk population. The thick disk appears to have very little gas, and so relatively few stars are formed there. It only has about 12 percent of the stellar density of the thin disk, but nonetheless the thick disk makes up about 10 percent of the mass of the galaxy. The orbits of the stars here take them above and below the plane of the thin disk for longer periods of time. The thick disk is about 2.5 times thicker than the thin disk, and may have formed either after a series of small interactions disturbed stars in the thin disk, or following an early large interaction, which disturbed the thin disk in the same way (fig. 29.1).

In the center of the Milky Way, and of most spiral galaxies, we have the bulge, which is a collection of generally older, and therefore much redder, stars than in the thin disk. These bulge stars are much less ordered than the stars in the disk, and are traveling in random orbits around the center of the galaxy, giving the bulge a not quite spherical shape. While there are young stars in the bulge, they are the exception rather than the rule. The fact that they are there, though, means that the population of stars in the bulge is somewhat younger than in the thick disk.

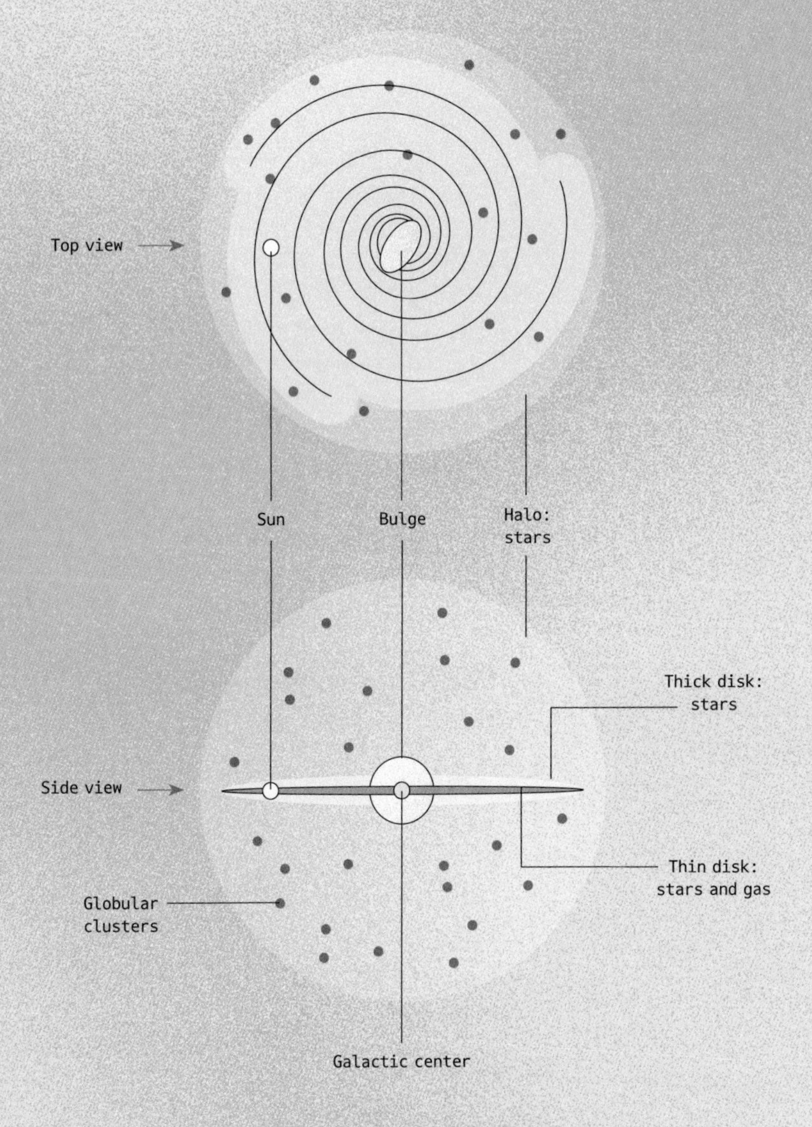

FIG. 29.1 Spiral galaxies can have the most complex structures of galaxy types, with two layers of a disk (thin and thick), a bulge, and a stellar halo.

Surrounding the entire dense part of the galaxy is the stellar halo. This is a much larger, faint, and diffuse assortment of very old stars, taking long paths around the galaxy. Some of these stars are so old that they can be considered stellar fossils—time capsules from an earlier epoch of the galaxy's history. Our best current model for how the halo is built is that it is the remnant of many past interactions between our galaxy and other galaxies it has encountered, especially small galaxies. As a smaller galaxy comes near to the Milky Way, the stars furthest from the center of the small galaxy are removed from their galaxy and pulled into orbit around the Milky Way instead. At first, these removals will be visible as streams around the Milky Way, but given enough time, the donated stars will spread out and scatter. Eventually they will disperse enough so that they simply surround the Milky Way in a faint haze. This stellar halo overall is roughly spherical in shape, and so faintly populated that it's only responsible for about 1 percent of our galaxy's mass (fig. 29.2).

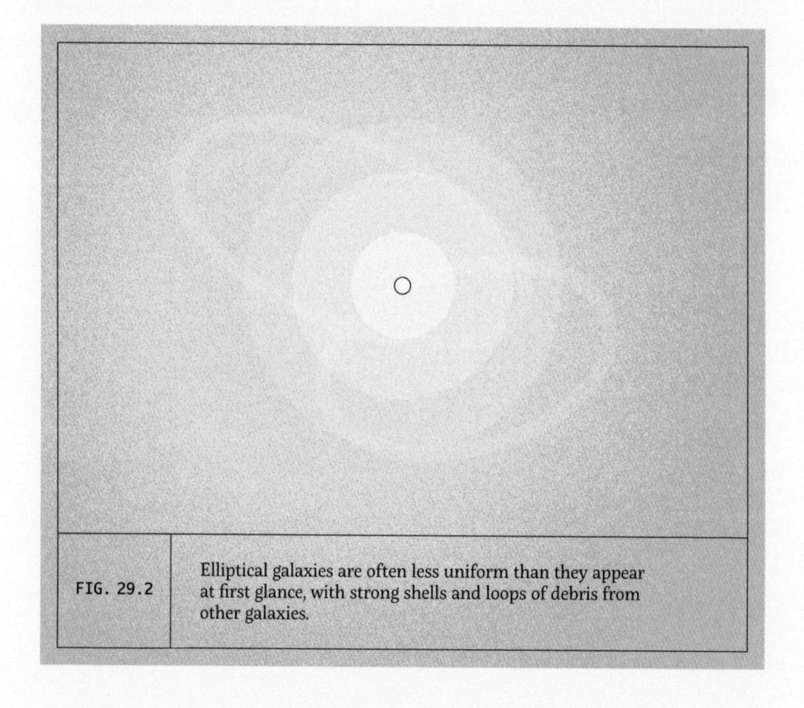

FIG. 29.2 Elliptical galaxies are often less uniform than they appear at first glance, with strong shells and loops of debris from other galaxies.

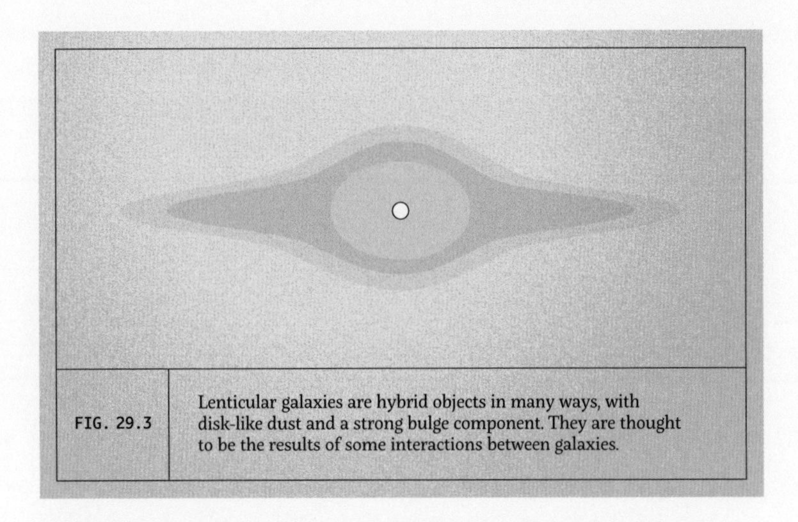

FIG. 29.3 Lenticular galaxies are hybrid objects in many ways, with disk-like dust and a strong bulge component. They are thought to be the results of some interactions between galaxies.

For any given galaxy, many of these structures may be present, but none are required. Spirals are sometimes identified with no substantial bulge component, which may indicate that they have lived in isolation with nothing to perturb the orbits of their stars. More typically, however, a spiral galaxy will have a noticeable, though not dominant, stellar bulge. On the other hand, elliptical galaxies are usually missing both possible disk components, and are made entirely of a galactic bulge and a stellar halo. These elliptical galaxies, when observed carefully, often show the results of a large number of previous interactions with other galaxies, showing streams and shells of stars discarded from a galaxy long since absorbed by the elliptical. But these trails of stars left behind are the hints we have to a tumultuous past.

In between the spiral and elliptical galaxies, there is a class of galaxy called lenticular, which has a large fraction of its mass in stars found in the bulge, but which still has a stellar disk. These galaxies often have quite a lot of dust, but not much in the way of spiral arms. They may represent incomplete transitions between spiral and elliptical objects. A spiral galaxy may have had its stellar disk disrupted into a bulge through an encounter with another galaxy, but the encounter was not sufficiently catastrophic to fully destroy the disk of the galaxy (fig. 29.3).

as a tracer of dark matter

Under standard circumstances, the orbits of any objects are well described by Kepler's laws, which say that there's a relationship between the mass of the objects orbiting each other and the length of time it takes for them to revolve around each other. And so, using this information, we can identify the mass of a star from the orbit of a planet around it, or the mass of a stellar system by watching how long it takes the two stars to orbit each other. But in some circumstances, the orbits of the stars hide a secret.

It then made sense to try this with the orbit of the stars around the center of the galaxies they inhabit. Rather than responding to the mass of individual objects, each star orbiting the galaxy responds to the collection of all the mass inside its orbit. To simplify, we often talk about this as being equivalent to orbiting a single object at the gravitational center of the galaxy with all the mass of the relevant stars held at that point, which is more correct as an assumption than it sounds like it should be.

The speed of the star's orbit depends on two things: the amount of mass inside the orbit, and how far away from the center of the galaxy the star is. For stars which are close to the center of the galaxy, the amount of mass within the orbit is small, and so the velocity will wind up being small, even though it's close. As the star gets further away from the center of the galaxy, the amount of mass it's orbiting increases, and so the velocity also increases.

At some point, though, we'd expect to run out of galaxy. There's an end to the luminous component of a galaxy, where stars become sparse. So as we get into the outer regions of a galaxy, we increase the distance from the center without increasing the amount of mass within a star's orbit. In that case, we expect the velocity of the star to decrease (fig. 30.1).

The first set of actual data to check whether this was how the stars were behaving arrived in 1962 with Vera Rubin, who determined that for our own galaxy, the stars were doing something else entirely. Their speeds had flattened out at a high value, rather than descending the way we would expect if we

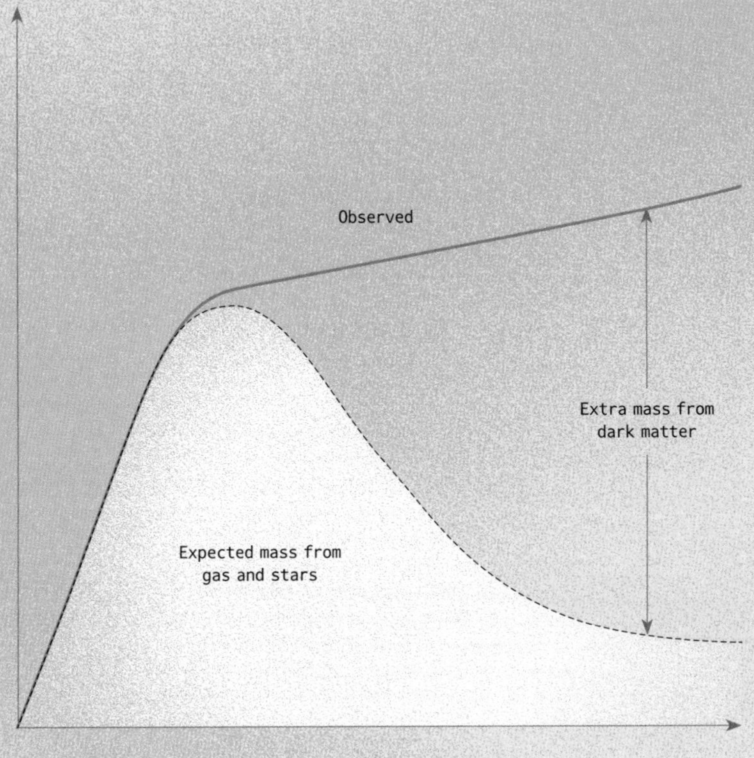

FIG. 30.1	The tell-tale sign of missing mass is the stars' maintaining high speeds as they orbit at increasingly large distances. The way to resolve this discrepancy was to realize we were missing some mass.

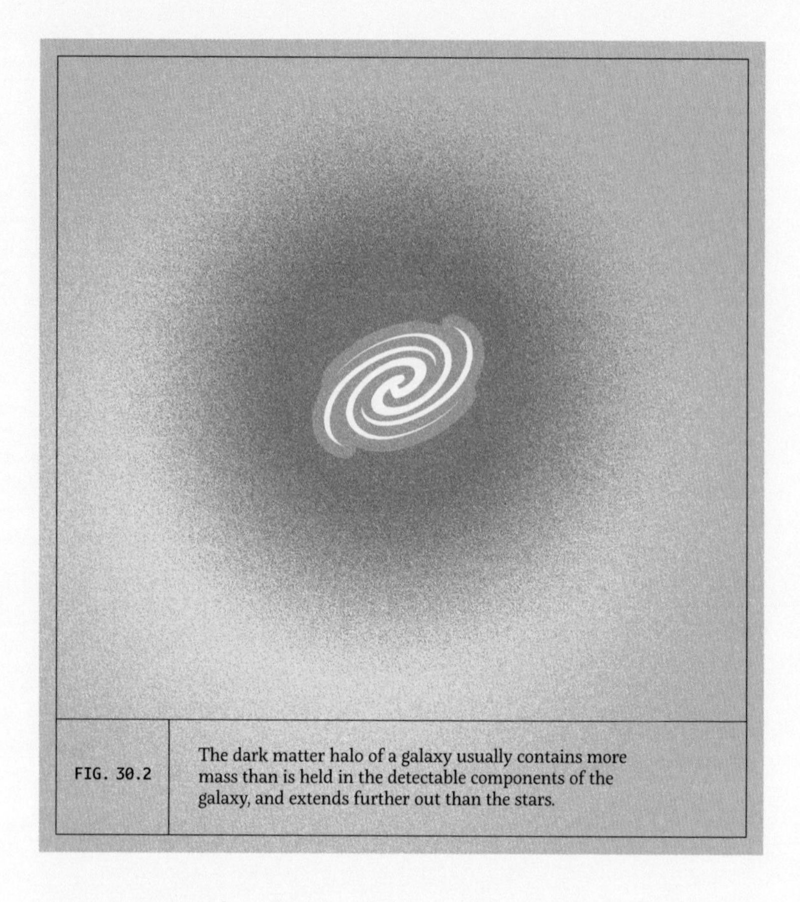

FIG. 30.2 The dark matter halo of a galaxy usually contains more mass than is held in the detectable components of the galaxy, and extends further out than the stars.

were measuring the velocities of the most distant stars, and the entirety of the mass of the galaxy were enclosed.

This created quite a puzzle, because our understanding of gravity had worked quite well right up until that point. So either our grasp of gravity was incorrect, which seemed unlikely given all the other tests it had passed, or our understanding of where the mass was in a galaxy was incorrect. To start to adjust where the mass in a galaxy is held, we had to expand our ideas past what was luminous and visible—which is the gas, stars, and dust. What we needed was a considerable source of mass that extended out past the stellar disk of our galaxy and was invisible to our telescopes.

This was one of the most solid pieces of evidence for what we now call dark matter—a form of matter which is not luminous, but gravitationally evident. Since then, Vera Rubin's discovery has been replicated on many other galaxies, both near and far, and every single one has behaved in the same way.

Galaxies, as traced by their stars, seem to be embedded in a spherical halo that extends far beyond the visible stellar disk, and which typically contains far more mass than can be explained simply by the mass in stars. Counting gravitationally rather than by the light from stars gives us two different masses for every galaxy: a stellar mass, which calculates the amount of mass we'd expect from its luminous pieces, and a dynamical mass, which counts the amount of mass traced by the orbits of its stars. This is the mass that the motions of the stars are responding to (fig. 30.2).

The dynamical mass is almost always substantially larger than the stellar mass. The exact difference varies a little from galaxy to galaxy, but in general, the dark matter component out-masses the luminous component. The total mass of a galaxy is considered dark-matter-dominated if it's more than 50 percent dark matter, and many galaxies are even more extreme. Some have more than 80 percent of their mass held in dark matter and only 20 percent in their detectable material. The stars, only a small fraction of the total mass of a galaxy, were able to reveal a hidden, but seemingly universal, component to the galaxies.

as part of dwarf galaxies

Within the smallest galaxies, the stars give us new hints toward understanding how a galaxy is made. Spiral and elliptical galaxies are the massive galaxies in our Universe, but if we look carefully in the area surrounding the Milky Way, we can find much smaller objects as well. The main difference is the number of stars hosted within them.

While massive galaxies have hundreds of billions of stars, these smaller objects, known as dwarf galaxies, usually hold no more than a few billion stars. Many of them resemble their massive counterparts, with dwarf elliptical galaxies being tiny analogs of an elliptical galaxy, and dwarf irregular galaxies mirroring the chaotic nature of the massive irregular galaxy. There is also a class of galaxy called an ultra-faint dwarf galaxy. These are faint, with only tens of thousands of stars, and they create a slight problem in how we draw a line between these and a globular cluster, which can also have tens of thousands of stars.

Globular clusters do tend to be denser and older objects than ultra-compact dwarf galaxies, but that's not a particularly satisfying way to distinguish between two objects that may well have fundamentally different pathways to forming. The current dividing line is the presence of dark matter. If the object is embedded in a dark matter halo, it's a galaxy. If there's no dark matter, it's a globular cluster. This works extremely well, because dwarf galaxies seem to be some of the most dark-matter-dominated objects we've ever seen (fig. 31.1).

Because dwarf galaxies are not very bright due to their small number of stars, most of the dwarf galaxies we know of are around the Milky Way, or around our nearest galactic neighbor, Andromeda. At greater and greater distances, they simply become too hard to see. While we are still discovering faint dwarf galaxies around the Milky Way, we've identified more than 50 dwarf galaxies in our little gravitational region, which far outnumbers the three massive galaxies present. The further away these dwarf galaxies are, the less information we have about them. This means that the dwarf galaxies we understand best are the ones closest to us,

Andromeda galaxy, M31

Milky Way

Triangulum galaxy, M33

4 million light years

FIG. 31.1	The Local Group is made of three large galaxies, but the rest of the population is a set of more than 50 dwarf galaxies, scattered around the massive galaxies.

where individual stars can be resolved, placed on the Hertzsprung-Russell diagram, and their histories and ages understood from the stars directly.

These diagrams indicate that dwarf galaxies also have complex histories, and the shapes of the galaxies reflect different histories. Dwarf dwarf ellipticals continue to mirror their massive counterparts. While they often show signs of distinct bursts of star formation in their pasts, they have left their star-forming days behind them, and don't have any of the bright blue stars with short lifetimes. Much like the giant ellipticals, the dwarf ellipticals have no easily detectable gas, and the redder stars that remain orbit the center of the dwarf elliptical in the same random, swarm-of-bees fashion. While it is tricky to estimate the dark matter content

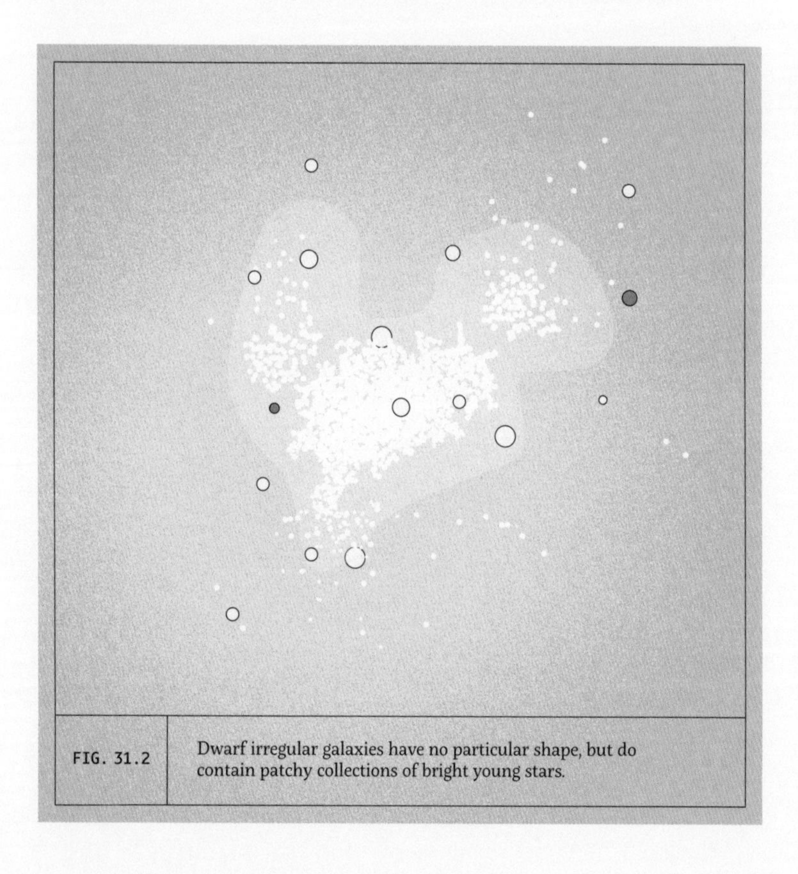

FIG. 31.2 Dwarf irregular galaxies have no particular shape, but do contain patchy collections of bright young stars.

in a dwarf elliptical because of the chaotic orbits, we can tell that there must be dark matter because these stars still orbit faster than we'd expect without dark matter being present (fig. 31.2).

Dwarf irregulars, by contrast, seem to have enough of the materials required to make stars—gas and dust—that their star formation is still ongoing, and so these galaxies are bluer and younger than the dwarf ellipticals. Some of these galaxies have been forming stars at a slow pace for a very long time, and others seem to be forming stars more dramatically now than they did in their pasts. The irregulars, true to their asymmetric shapes, do not show the kind of coherent rotation we can expect of a spiral galaxy, and so undertaking the rotation curve work to estimate their dark matter components is much trickier than it would be if there was more ordered rotation. We can, however, use the same methods as for the dwarf ellipticals, and these analyses show that the stars and gas of the dwarf irregular are also embedded in a much larger dark matter halo.

These galaxies sometimes seem to be strongly affected by being close to a galaxy much larger than themselves. Some of these dwarf galaxies show signs of being tidally stretched out by their massive companions, and a few of them are suspected to have once been much larger than we currently see them, their outer layers having been lost to the stellar halo of the galaxy they orbit. This may not be universally true of all dwarf galaxies, but the set we can see clearly is by definition those that are close to our massive galaxy, so we're biased to spotting these partially disrupted systems.

PLATE 9

A translucent dwarf irregular galaxy

NGC 5477 is a chaotic mess of stars and the glow of hydrogen gas, but so faintly populated with stars that you can see through it, to the glow of distant galaxies.

by its metals

While we know that the stars are overwhelmingly made of hydrogen, there are usually other elements mixed in with that in small quantities, and there is much to be learned from these trace quantities of heavier elements.

Astrophysically, any element heavier than helium is a metal, which usually upsets the chemists. Naming conventions aside, this split between hydrogen and helium and everything else is sensible, because everything heavier than helium in our Universe was created in a star. These metals, once created, are distributed energetically back into the galaxy by the supernova (or planetary nebula) of the star. A lot of this returned gas is still hydrogen, as the outermost layers of massive stars are still hydrogen-rich, and are first to be lost to the galaxy again. However, mixed in with the hydrogen will be the heavier elements formed within the fusion shells or in the shock front of the supernova explosion. When the next round of stars forms in that region of the galaxy, it will be collapsing down from a gas cloud that's pre-enriched with metals from this previous generation of stars (fig. 32.1).

This metal enrichment gives us a particularly useful new metric for measuring the relative ages of the stars in a galaxy, especially among the long-lived stars. Relatively low mass stars are formed with every generation of star formation in a galaxy, and because their lifetimes are long, they will remain on the main sequence for a long time. The absence of blue stars can tell us that there hasn't been a lot of star formation in that area "recently," but it can't tell us how many of those bursts there have been. The metals can.

Recently formed stars will have a relatively large amount of metals within them, because the gas has been recycled for more than 10 billion years, in and out of stars as they explode, and so the metal content of that gas will have increased with every single generation of massive stars that undergo a supernova. Stars which formed a long time ago, however, will have fewer metals in their atmospheres, reflecting the smaller number of stellar generations that preceded them. If you can measure the proportions of metals in a star—known as its metallicity—then you can start to order

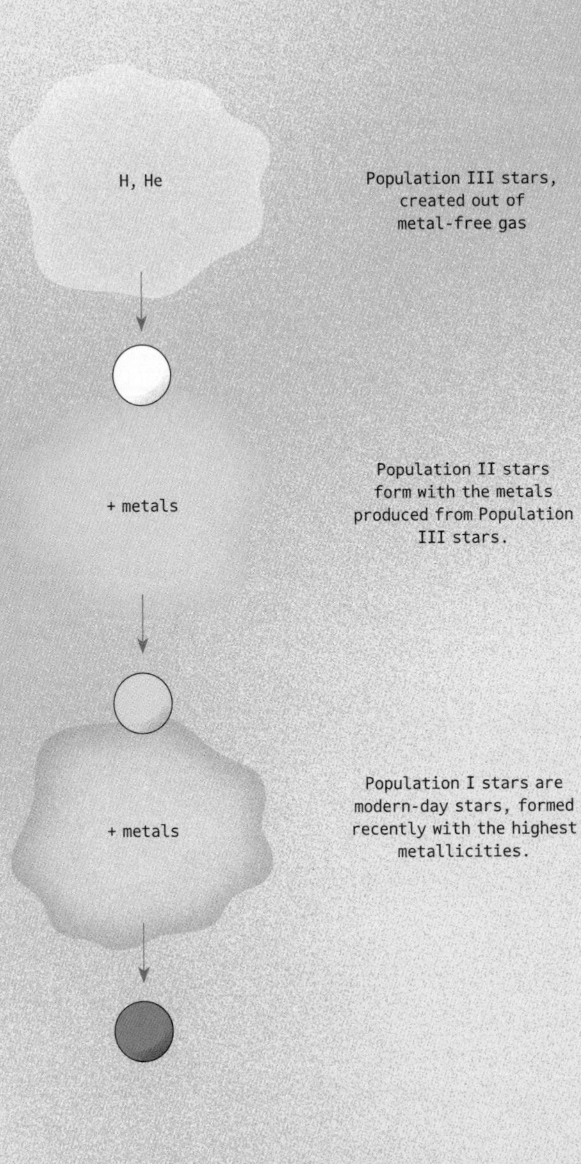

FIG. 32.1	The metals present in the stars as we see them today have been built up over successive generations of stars forming and dying over cosmic time.

the stars by how long ago they formed, even when they are all the same color (fig. 32.2).

Broadly, we can classify stars into three populations, based on their metallicity. Population I stars are those that are forming in the modern-day Universe. These are metal-rich stars, ranging from somewhat less to several times more metal-rich than our Sun, and we'd expect to find them in the active star-forming regions of our galaxy and other galaxies. They are found predominantly in the thin disks of galaxies.

Population II stars are dramatically less metal-rich than their Population I counterparts, and reflect much older stellar populations. Globular clusters typically are a good reservoir of these stars, and in fact their low metallicity is part of the evidence for

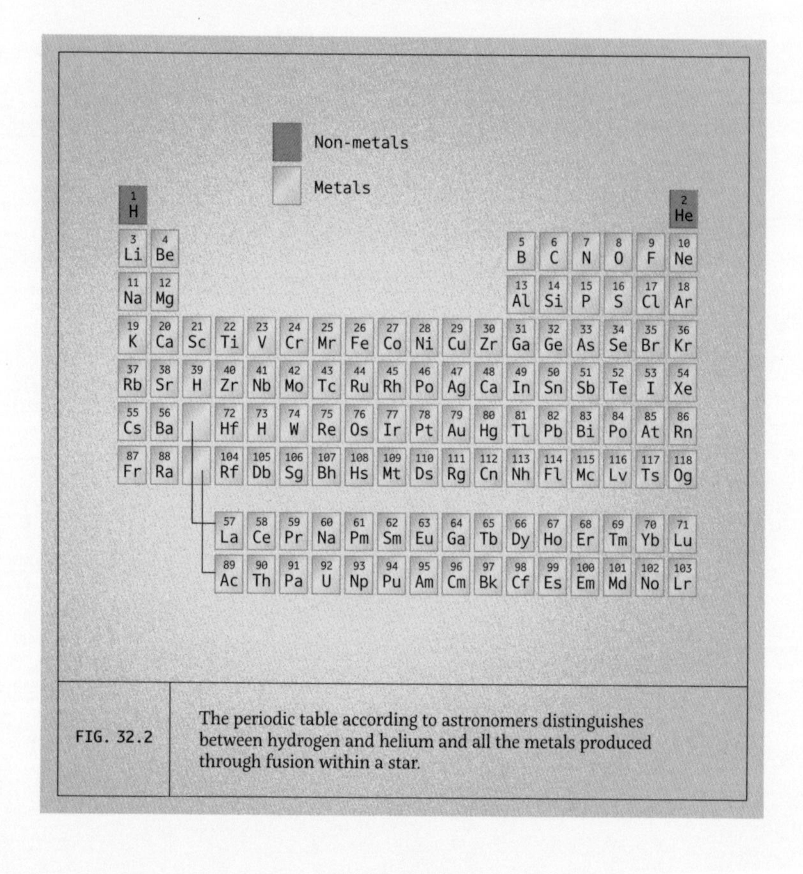

FIG. 32.2 The periodic table according to astronomers distinguishes between hydrogen and helium and all the metals produced through fusion within a star.

how old they must be. Dividing stars into Population I and II has been helpful for learning about the structures of galaxies, since many of the components of a galaxy show different metallicities, and this teaches us about the ages of these structures.

However, even the Population II stars have metals present in their atmospheres, and so there should have been a generation of stars prior to the formation of Population II, old as they are. So far none of these very early stars has been found, but they have been given the term Population III while the hunt is underway. Population III stars should, in principle, be metal-free, since our models of the Universe do not include any way to create iron other than through a star. And we would expect, if Population III stars formed like stars today, that there should be some low mass Population III stars hanging about in our galaxy, remnants from the earliest Universe. However, despite quite a bit of hunting, no probable candidates have ever been detected. This lack of ancient metal-free stars has been a puzzle, and so perhaps Population III stars didn't form like stars today. One suggestion is that all the first generation of stars were massive, short-lived stars. If that were the case, then we wouldn't expect to see any long-lived low mass stars left over from this early generation. This could explain how we got enough metals in the Population II stars to match what we see, and also account for the missing first generation of stars in our stellar neighborhood.

as proof of massive black holes

We can also watch the stars trace out the otherwise invisible gravitational field of a massive black hole. Black holes in isolation are extremely difficult to spot. The fastest and easiest way to spot them is with the presence of superheated material in an accretion disk. But there are many instances where that material isn't present, and the black hole fades from visibility. If the black hole is small enough, we may simply miss it. But if it's sufficiently large, we can spot it through other methods.

It's thought that at the center of every galaxy lies a tremendous black hole, millions of times more massive than our Sun. For the Milky Way, this black hole is an object known as Sagittarius A* (pronounced A-star). This class of black hole, to be distinguished from a stellar mass black hole formed after the supernova of a massive star, is called a supermassive black hole. The black hole in the center of our own galaxy seems to check in at some 4.3 million times the mass of the Sun, all packed within a volume that would fit within the orbit of Mercury around the Sun.

We have had plenty of evidence in recent years that this object is truly a supermassive black hole, including the fantastic image of Sagittarius A* from the Event Horizon Telescope in 2022, which showed a direct view of the shadow of the black hole for the first time. But prior to 2022, we had another line of evidence pointing strongly to a black hole, and that was from the dedication and work over multiple decades from two independent teams, carefully mapping out the orbits of the stars closest to the galactic center (fig. 33.1).

There is one star in particular that's been exceptionally helpful for this work, nicknamed S2, which completes one orbit every sixteen years. Scientists have mapped this out for more than one complete orbit. In order to be orbiting that quickly, its companion object should be extremely massive, but there's nothing luminous within the orbit of S2. The lack of a bright object for S2 to orbit was the first hint that there might be a black hole there. The second hint was just how massive the object had to be. Purely from the orbits, studies predicted that the invisible object would have to contain some 4.3 million solar masses, far beyond what we'd expect from a single, luminous object.

FIG. 33.1 The orbits of the stars immediately surrounding the black hole at the center of our galaxy were some of the earliest strong evidence that a supermassive black hole lurked at its core.

As a final piece to this puzzle, in 2018, S2 completed its closest passage to the massive object it orbits, passing a mere 120 AU (the distance between Earth and our Sun) from the invisible gravitational heavyweight. It whizzed past at a tremendous pace: 4,750 miles per second (7,650 km/s), or about 17 million miles per hour (28 million km/h). At that speed, one could circle the Earth in five and a quarter seconds. If the object were made of material like even the densest stellar remnants it should have taken up so much volume that S2 should have encountered it, but instead it carried on unperturbed. The simplest explanation for such

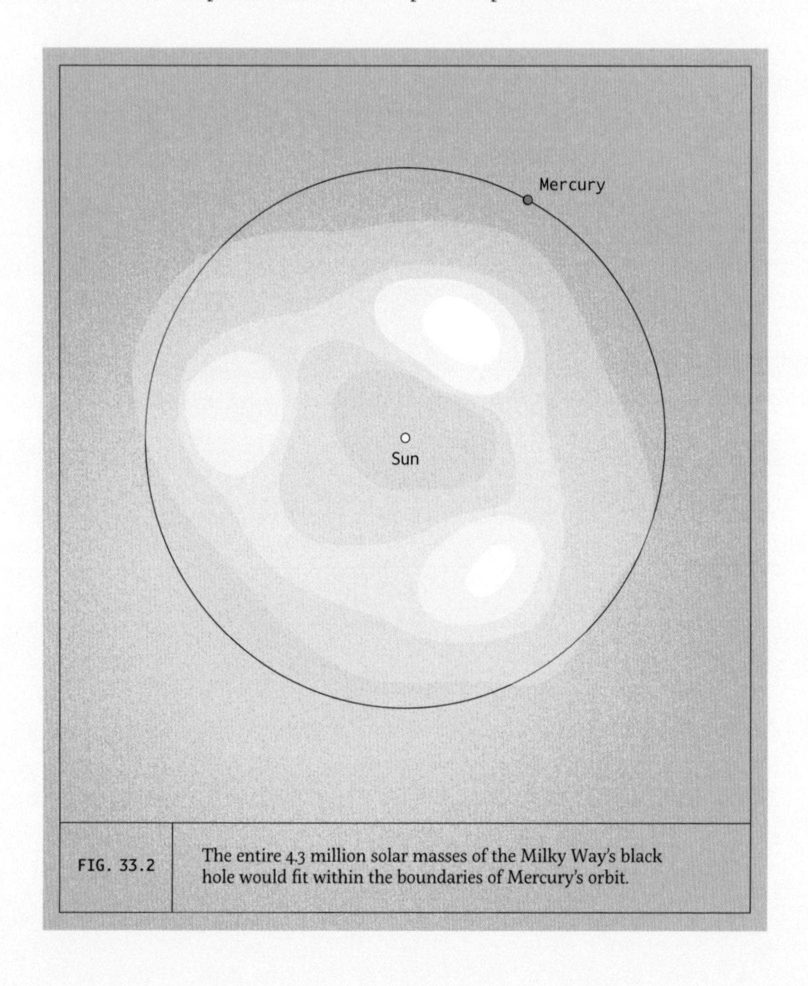

FIG. 33.2 The entire 4.3 million solar masses of the Milky Way's black hole would fit within the boundaries of Mercury's orbit.

a massive object, packed so densely and orbited so rapidly by S2, is that of a supermassive black hole. With the imaging of the black hole in 2022, all remaining doubts were dispelled (fig. 33.2).

Recently, a similar method has been used to hunt for black holes in other objects, using a twenty-year baseline of data taken for other purposes by the Hubble Space Telescope. The unusual object Omega Centauri exists on the boundary between dwarf galaxy and globular cluster, but close examinations of it have offered a potential solution to this classification difficulty. Omega Centauri may be the remains of a much larger galaxy, with everything except the concentrated nucleus already lost to space. This has left an object with the densities of a globular cluster—as indeed it was classified when first discovered—but with too much mass to match well with the rest of the globular clusters. Recent observations have indicated that it may also have a black hole in its center, using the same kind of stellar tracing that revealed the Milky Way's own black hole. If confirmed, the mass of this black hole would fall in between that of a Milky Way-like supermassive black hole and a stellar mass black hole. This is called an intermediate mass black hole, and evidence for any of them has been hard to find thus far. However, the tracing of a set of stars in the center of Omega Centauri showed that they were moving far too fast to be gravitationally bound to Omega Centauri unless a black hole of at least 8,200 times the mass of the Sun was present, and potentially up to 40,000 solar masses.

as a distance marker

The stars have also been able to mark the distances between us and themselves. There is a particular type of variable star now known as a classical Cepheid variable, after one of the first identified stars of this form, which was found in the constellation of Cepheus. These stars, generally more massive than our Sun, regularly and repeatedly pulse in brightness with a distinctive sharp rise and slower fall pattern, reflecting a physical change in size.

Between 1908 and 1912, the astronomer Henrietta Swan Leavitt published two works analyzing these Cepheids in one of the Milky Way's small companion galaxies, the Small Magellanic Cloud. The first of these identified 1,777 variable stars, and the second published what is sometimes now known as Leavitt's Law: that there is a distinct relationship between the length of time between bright pulses of a variable star and the intrinsic brightness of the star.

All the stars in the Small Magellanic Cloud are at roughly the same distance from Earth, since they're all held within the same object. So, to see a relationship between the apparent brightness of the Cepheid and the frequency of its flickering meant that these could become a "standard candle" in the Universe. Standard candles are anything where we know how bright the object should be. If we can compare our understanding of how bright it should be with how bright it actually appears in the sky, we can figure out how far away it is (fig. 34.1).

More distant objects are always fainter to our observatories on Earth, and there's a straightforward relationship between the distance of an object and how much fainter it is. Double the distance, and the object will appear four times fainter. Triple the distance, and the object will be nine times fainter. Quadruple the distance, and it will be sixteen times fainter in the sky. It can be quite tricky with just a single image to distinguish between a star that's intrinsically faint and one that's simply far away. In the case of Cepheid variable stars, we have an extra piece of information: how much time passes between bright peaks.

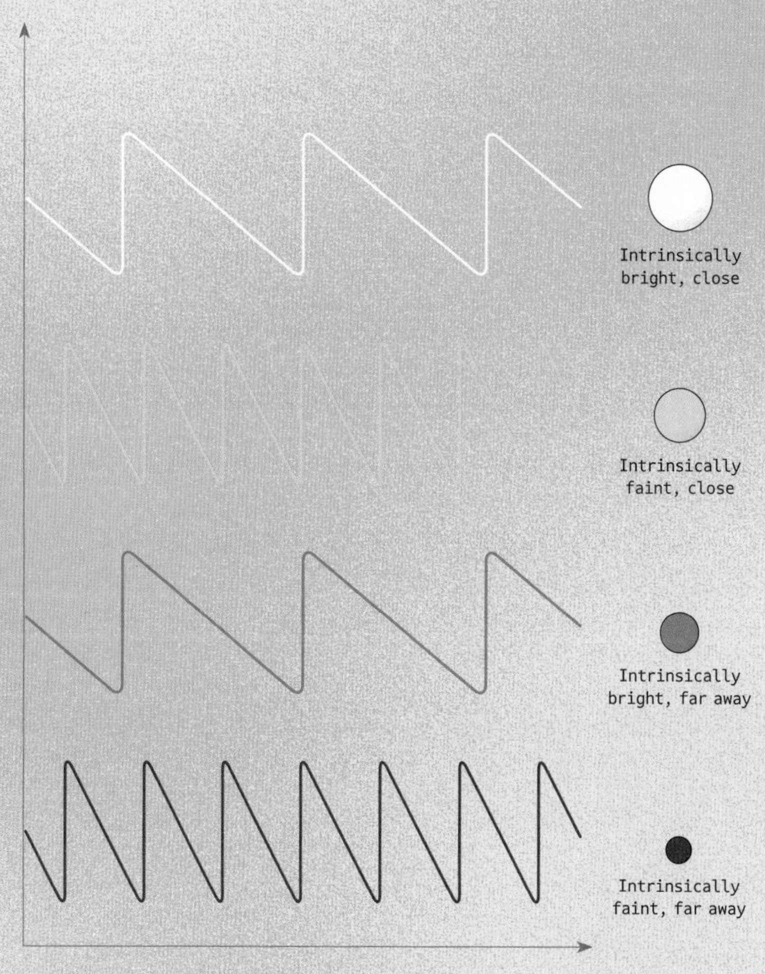

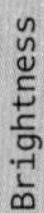

FIG. 34.1	The sawtooth typical curve of a Cepheid variable star is more widely spaced the brighter the star is intrinsically, which allows us to measure distances to faraway stars.

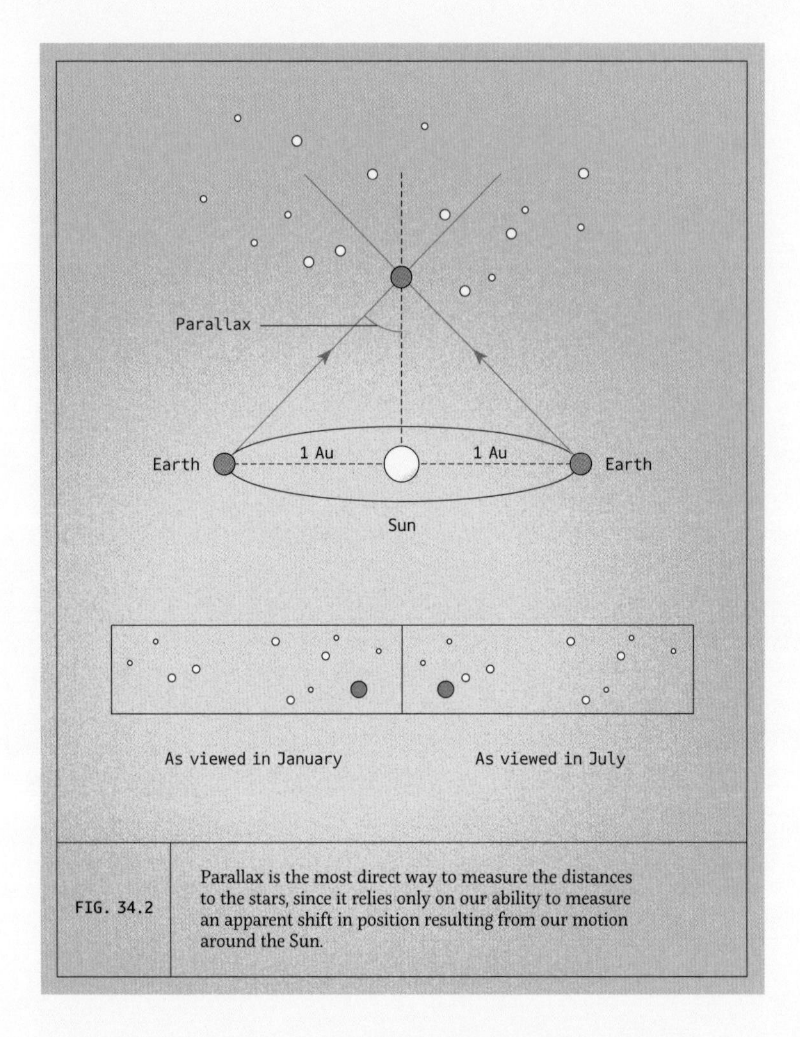

FIG. 34.2 Parallax is the most direct way to measure the distances to the stars, since it relies only on our ability to measure an apparent shift in position resulting from our motion around the Sun.

What Henrietta Leavitt found was that the brightest Cepheids had the longest periods. And this meant that if you could observe the flickering of a Cepheid, you knew whether it should be bright or faint, intrinsically. If you could compare that with how bright or faint it actually appeared, you could figure out the relative distance.

To translate this into actual distances instead of just relative ones, we need to find the period of a Cepheid that's close enough to Earth that we can measure the distance in a different way:

through a parallax. Parallax is the phenomenon that we see when moving quickly—say in a car or train—and objects close to us appear to move much more quickly past us than objects in the far distance, which may appear stationary. To do this with the stars, we need to move as much as possible, and then we can look to see if our nearest stars have moved with respect to much more distant stars, which will appear as stationary as a background mountain range. The furthest we can move on Earth is to wait six months for our planet to carry us halfway around our orbit of the Sun. Since we know the distance between the Earth and the Sun very well, we know exactly how far half a year takes us. With that information, we can make a giant, skinny triangle using the angle we measure for how far a star appears to move in the sky to get a clean measure of the distance to that star (the long edge of our triangle). Even then, sensitive instruments are required. The Hipparcos satellite, which was in orbit around the Earth from 1989 to 1993, was the first space mission to undertake this task, and it logged distances to about 640,000 stars. The Gaia satellite has since superseded this with distances to 1.7 billion stars, but it uses exactly the same principle (fig. 34.2).

If we can get a parallax to a few Cepheid variable stars, then we have a direct measurement of the distance, and a measure of the brightness, and we can tie that to the length of time a Cepheid takes to flicker. And we can indeed directly measure the distances to Cepheid variable stars within the Milky Way using parallax. In fact, the Gaia Data Release 3 has 15,000 of them. This is all we need. With that ability to benchmark, we can measure the distances to stars far beyond where we can use parallax by instead measuring the length of time it takes a star to flicker. It was through a Cepheid that we proved Andromeda is definitively not part of the Milky Way, but millions of light years distant.

[THIRTY-FIVE] Know a star...

at cosmic noon

When we observe galaxies by looking at their stars, we see them at one moment in time—however long ago the light from the stars left them to travel the vast distances between us and arrive to Earth. For any galaxy outside our own, this travel time will be millions of years, with more distant galaxies showing us how they looked potentially billions of years in the past, when the stars released their light.

We can learn a lot about galaxies simply from comparing them at different distances, but sometimes a more in-depth approach is desired. In those cases, we might take a galaxy near to us and try to untangle what it might have gone through in the billions of years before its light reached us, reconstructing the galaxy's history. Usually, this means constructing the star formation history of a galaxy, or a timeline of when the galaxy formed the stars we see. This allows us to dig beyond just a map of what stars exist in the galaxy now, and create a model of what it might have looked like in the past. This, in turn, allows us to understand how galaxies were assembled into the structures we see today, by learning how they built up their mass in stars (fig. 35.1).

The Milky Way is one of the easier galaxies to undertake this task for, since we have the most detailed view of the stars that make up our own galaxy. In order to figure out when the stars in the Milky Way were formed, some studies have used surveys to identify those stars that are undergoing hydrogen shell fusion after their departure from the main sequence. These stars are particularly helpful for building up a timeline of events because models can use their brightness to determine an age. Because the hydrogen shell fusion period for any given star is relatively short, the models are able to be reasonably accurate in their determinations of how long ago the star itself formed.

There will be stars of all kinds of ages in this phase. More massive stars will be relatively young, but ancient low mass stars can simultaneously be in this evolutionary period. These aged stars will have delayed their arrival to the hydrogen shell fusion phase by spending much longer on the main sequence.

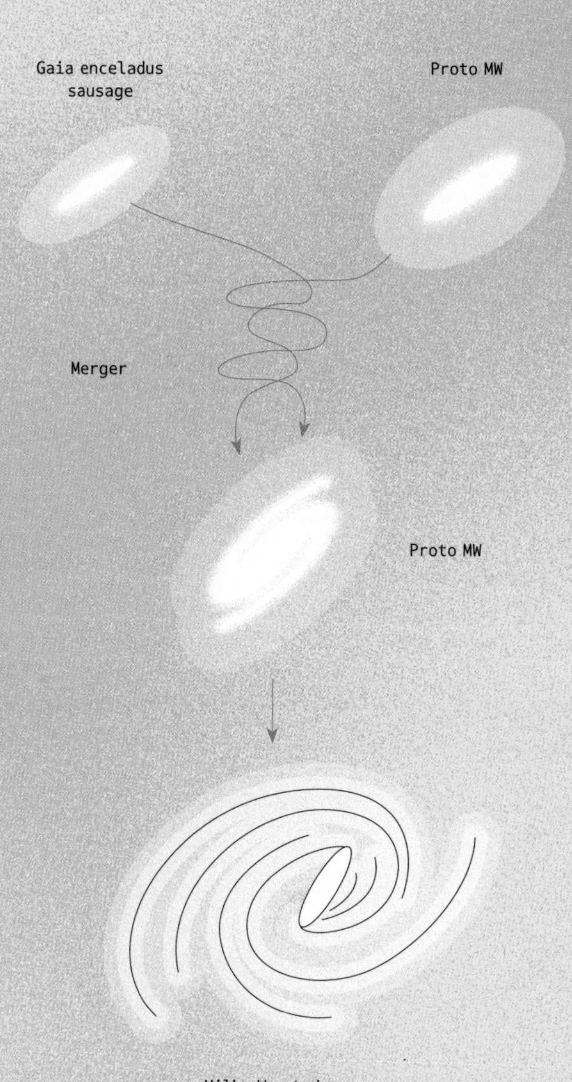

FIG. 35.1	The Milky Way seems to have undergone an abrupt burst of star formation some 11 billion years ago, potentially triggered by encountering another galaxy.

One study found that the Milky Way must have formed a large number of its stars a long time ago: somewhere around 11 billion years in the past. The thick disk of the galaxy, which hosts older stars than the thin disk, seems to have begun to form some 13 billion years back, with the majority of its stars forming in a burst around 11 billion years ago. It continuously declined in star formation until about 8 billion years ago, when stars broadly stopped forming in the thick disk. This ending of star formation corresponds to the relatively redder and aged color we see in it today. The tremendous

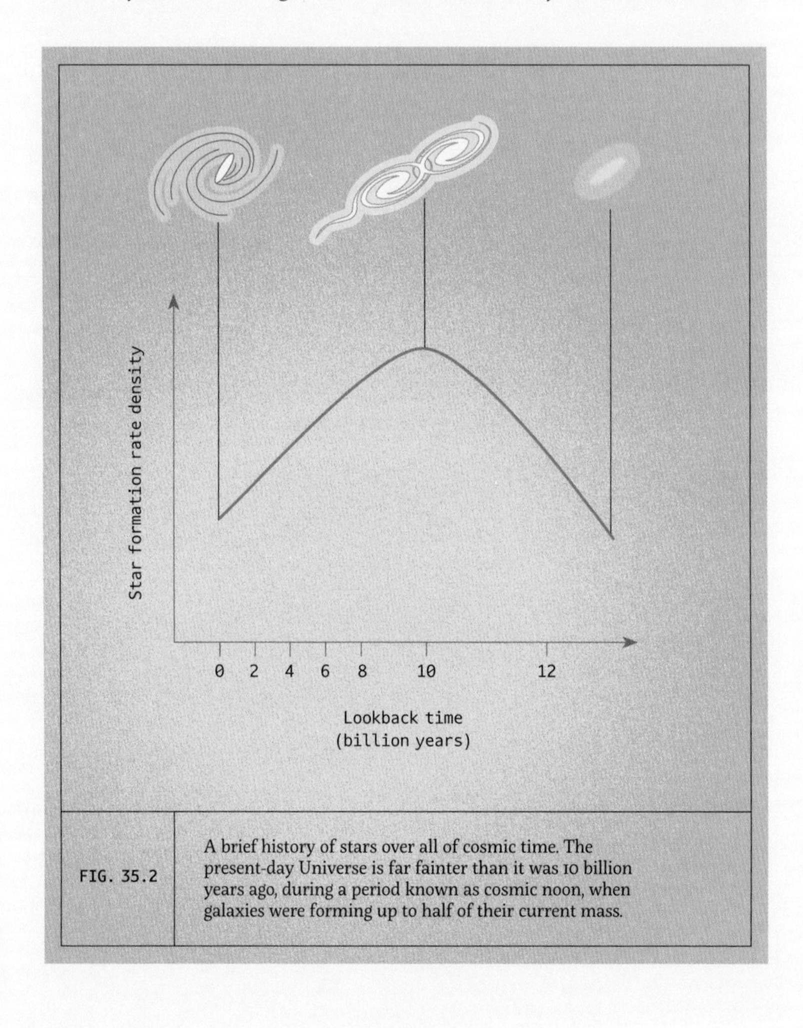

FIG. 35.2 A brief history of stars over all of cosmic time. The present-day Universe is far fainter than it was 10 billion years ago, during a period known as cosmic noon, when galaxies were forming up to half of their current mass.

amount of star formation 11 billion years ago is thought to correspond to an ancient interaction our Milky Way must have had, early in its life.

A broader census of other galaxies reveals that the Milky Way is not alone in having an early burst of star formation. Our galaxy is a little precocious compared to the average, but on the whole, star formation in galaxies was at its most vigorous 10 billion years ago, during a time nicknamed "cosmic noon." In this period the typical galaxy shone brightly, with hot, blue stars churned out of massive gas clouds. These blue stars glow brightly in the ultraviolet, and space-based telescopes carefully collecting the faint light shining through the cosmic distances have been able to capture the light emitted by these stars over many eras. As our telescopes improve, and our ability to capture light from more and more distant galaxies increases, we've built up our understanding as well. Cosmic noon was the Universe's brightest epoch (fig. 35.2).

These temporarily brilliant stars would have been accompanied by the formation of countless smaller, redder stars, whose lifespans continue to the present day. It's these Sun-like (and fainter) stars that truly built up the mass of the galaxies. It's estimated that perhaps half of the mass in a galaxy as we see it today may have been built up over the course of only 3.5 billion years, starting some 11.5 billion years ago and continuing until about 8 billion years ago—roughly the same time frame when the Milky Way's thick disk was forming its own stars.

in a galaxy's collisions

Extreme situations are also a fun testbed of star formation. Most of the time, galaxies live in relative isolation, and whatever happens internally in a galaxy is a product of the interactions between the reservoirs of gas, dust, and stars that exist within it. However, it's thought that most galaxies have gone through periods when this was not the case. At these times they have swung too close to another galaxy to have their behavior only controlled by internal mechanisms.

If you put two galaxies close enough to each other, each will experience an asymmetrical pull from the gravity of the other galaxy. This has the impact of destabilizing an otherwise placid structure, stretching out the galaxy through a gravitational tidal force. Galaxies, unlike smaller objects, have no particular structural integrity. Each star is a distinct, free-floating object, and the galaxy is simply made up of a large number of objects controlled by their collective gravity.

As a result, it's quite easy to stretch out a galaxy. The orbit of any given star responds to the distribution of mass around it. If we change that distribution by introducing another galaxy to the equation, we will change the orbit of the star. If we just had two galaxies next to each other, the stars on the near sides would stretch out toward each other, but the situation is more complex, because both galaxies are moving relative to each other. This means that the gravitational perturbance to each galaxy changes as a function of time, and the way that the stars orbit is continuously disturbed.

In general what we see is the development of tidal arms, peeling away the near and far sides of the galaxy, flung outward in response to the closest passage of the companion galaxy, when the masses of the two galaxies were closest together and gravity was strongest between the two objects (fig. 36.1).

If the merging galaxies are spirals, the rattling of the galaxy does more than shake up the orbits of the stars. Gas clouds are also disturbed by the gravitational arrival, and unlike a star, the gas within a cloud can collide with itself. The gas within a galaxy tends to shift inward slightly as a result, and the central region of the galaxy will fill with enough inflowing gas to create a rich reservoir

FIG. 36.1 The process of two galaxies merging together is a rapid and effective way for galaxies to grow in mass, and has the side effect of scrambling stellar orbits and forming many new stars.

of dense gas—ideal for the formation of a large number of new stars. The galaxy, as a result, will flare into light in its central region: bright blue stars will be visible.

Depending on how dramatic the interaction between the two galaxies is, the number of stars formed in this central starburst will also change. Things that increase the gravitational perturbation will increase the strength of the starburst, so the closer the galaxies are, in general, the more stars are formed. And the closer in mass the two galaxies are, the more stars are formed.

It's possible for galaxies to have a single close passage and be moving so fast that they swing past each other and outward, never to encounter each other again. These are called flybys, and while they can trigger changes in the stars' orbits and cause a burst of star formation, they're not nearly as dramatic as what happens when the galaxies have slower relative speeds (fig. 36.2).

In those cases, the galaxies are gravitationally caught. Once they swing past for the first time, they will weave an intricate path around each other, but they are gravitationally bound to fall together into a single object, given enough time. "Enough time" for a galaxy often means several billion years, but some galaxies have more direct paths to merging. In these states of merging into a single

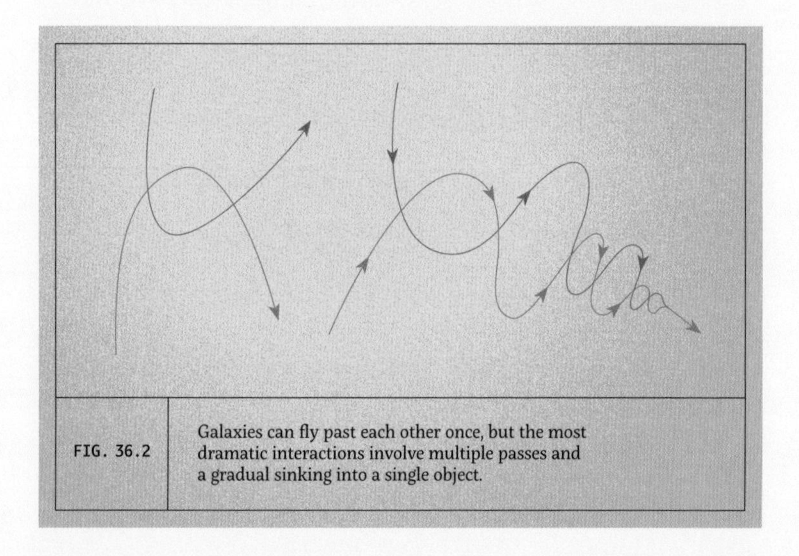

FIG. 36.2 Galaxies can fly past each other once, but the most dramatic interactions involve multiple passes and a gradual sinking into a single object.

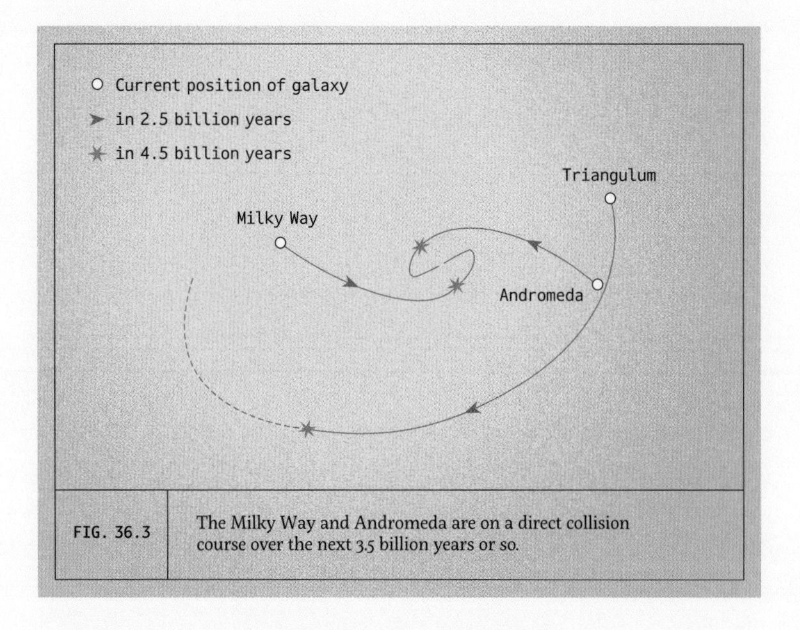

FIG. 36.3 The Milky Way and Andromeda are on a direct collision course over the next 3.5 billion years or so.

object, we can sometimes catch a galaxy forming more than ten times the usual number of stars for its mass. These final stages, where two roughly equal mass galaxies merge, will also profoundly change the shape of the galaxy, scrambling the orbits of the existing stars and tearing through the gas present in both galaxies in order to form new stars. In extreme cases, these interactions can destroy the disk of a spiral galaxy, consuming all of its gas and converting a spiral into an elliptical galaxy (fig. 36.3).

These mergers are considered to be the main mechanism by which galaxies have built their mass up to the hundreds of billions of stars that we see today, and are also the way that elliptical galaxies have been built. Interactions with less massive galaxies have smaller impacts on the massive galaxy, but can nonetheless leave streamers of stars pulled away from the smaller object. The Milky Way has many such streamers from dwarf galaxies encountering our much larger home galaxy. A massive collision is in our future, however: Andromeda and the Milky Way are falling together, and should begin this chaotic process in another 3 to 4 billion years.

PLATE 10

A peculiar chaos

UGC 8335 is a set of interacting galaxies in the constellation of Ursa Major, showing a bridge of material connecting the two systems, and tidal "tails" of material flung away from what was once two separate galaxies, now on their way to merge into a single galaxy.

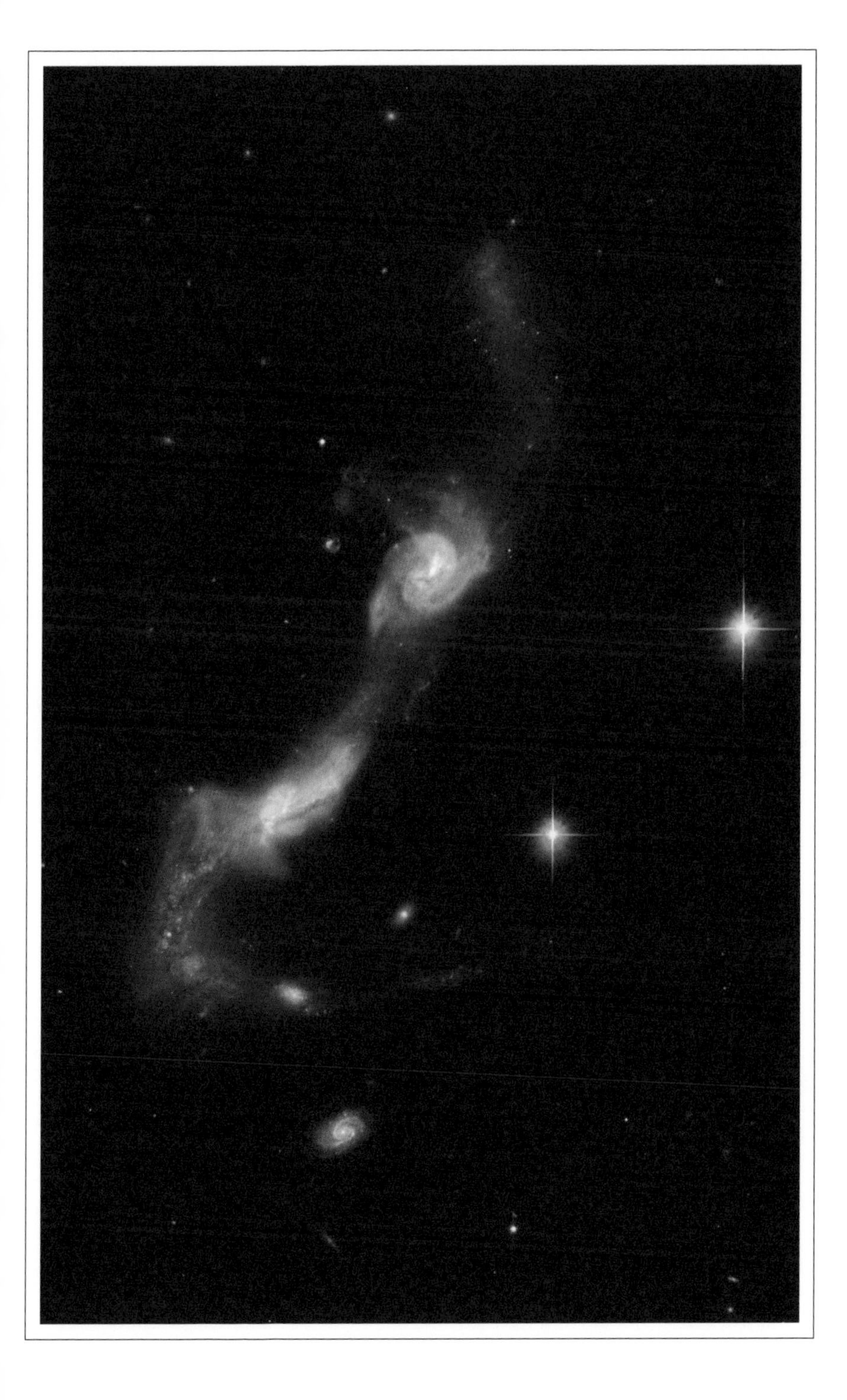

as a type 1a supernova

In particularly dramatic situations, we can watch a star self-destruct, leaving nothing behind. This event is a catastrophic explosion, and requires quite a particular set of circumstances in order to occur. We begin with a white dwarf in a binary star system, either with a massive companion star, or with another white dwarf. If the companion is a massive star, we have the same setup as we needed in order to generate a regular nova event: the repeated surface explosions when material siphoned off of the massive star accumulates on the surface of the white dwarf and becomes hot enough to start fusion.

However, there are limits to this situation. White dwarfs have an upper mass limit beyond which they are unstable, and it strikes at the specific mass of 1.44 times the mass of the Sun. This is called the Chandrasekhar limit. Beyond this mass, the pressure from electron degeneracy is no longer able to support the white dwarf against gravity, and the entire object will become unstable. This will very rapidly trigger a new kind of supernova explosion: the type 1a.

The distinction between when a white dwarf undergoes a nova vs. detonating as a supernova seems to be the rate at which new material is unloaded onto its surface. If the material accretes slowly, then the white dwarf is much more likely to detonate in a nova explosion before it gathers so much mass that it reaches the Chandrasekhar limit. If the mass accumulates quickly, however, then the white dwarf is likely to reach that limit (fig. 37.1).

Once the mass limit is reached, the white dwarf only has a few more seconds to survive. The temperatures inside it will increase to the point that the carbon in the white dwarf will begin to fuse into heavier elements, and this fusion will very rapidly consume the entire stellar remnant, releasing so much energy so abruptly that the white dwarf will self-destruct, detonating itself in one final burst of energy release. These events are so luminous that they can temporarily outshine the rest of the stars in their galaxy, and because the white dwarf flings itself apart, there is absolutely nothing left behind once the explosion fades, apart from a bubble of gas. Unlike a core-collapse supernova, there is no neutron star

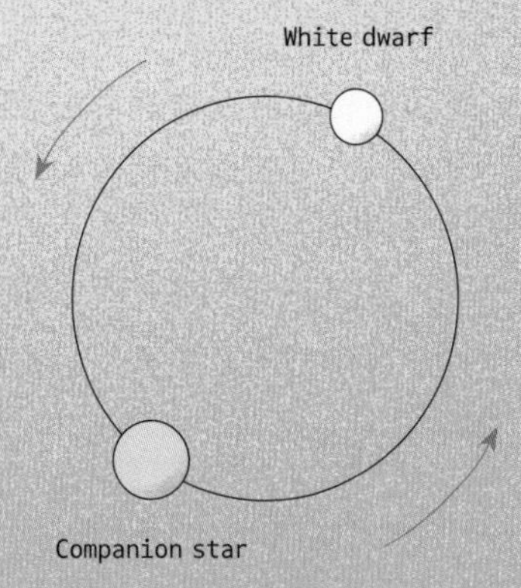

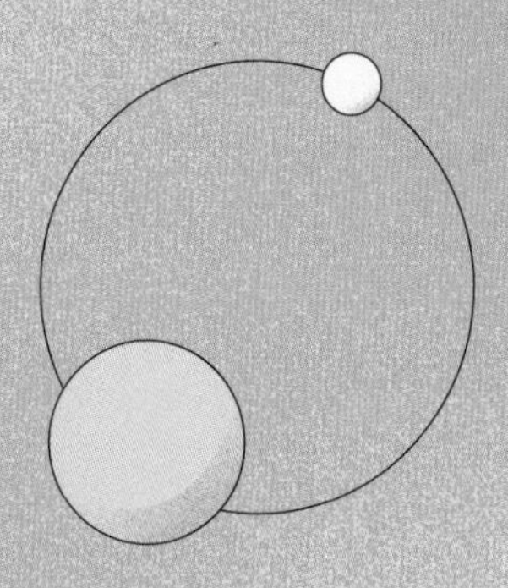

Swells through time

The intense gravity of the white dwarf pulls material from the companion star, causing it to gain mass

The white dwarf detonates once it surpasses its stable mass limit. This can sometimes eject the companion star

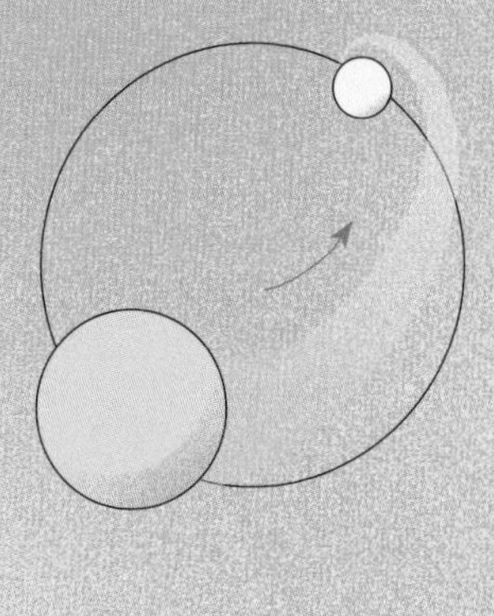

FIG. 37.1	Type Ia supernovae are complete detonations of a white dwarf star that has gained too much mass to be stable. They are extremely bright, which allows us to see them far beyond where Cepheids can be used to trace out a distance measurement.

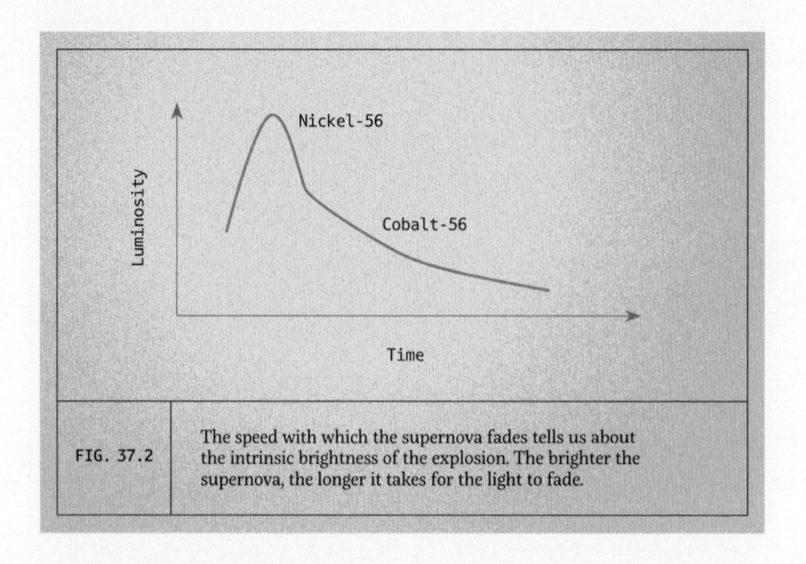

FIG. 37.2 The speed with which the supernova fades tells us about the intrinsic brightness of the explosion. The brighter the supernova, the longer it takes for the light to fade.

or black hole that remains after the detonation of the white dwarf. The companion star may partially survive, though we'd expect a large fraction of its outer layers to be stripped away as the shock wave of the supernova blows past (fig. 37.2).

This form of supernova—where a companion star causes a white dwarf to exceed its stable mass—is a particularly useful form of outburst, because the explosion is coming from the same type of object, at close to the same mass, every single time. And so, much like the Cepheid variable stars, this means that intrinsically, the objects should be undergoing roughly the same detonation. Some early work showed that the intrinsic brightness of the supernova is correlated with how slowly it fades from its peak brightness. The fainter the supernova, the faster it fades. This means that supernovae Ia can be used as a standard candle, much like the Cepheids. By comparing the predicted absolute magnitude of the supernova with the brightness we observe, we can figure out how far away the supernova must have taken place.

In contrast to the Cepheids, which can only be measured close enough to Earth that individual stars can be observed, the supernovae Ia allow our distance measurements to stretch much further. The profound brightness of supernovae explosions means that we can detect them in very distant galaxies, even when we have trouble spotting the galaxy itself (fig. 37.3).

There is one other pathway to a supernova Ia, and that is when the white dwarf has another white dwarf as a companion. In this case, material isn't slowly pulled off of the companion star—it's the entirety of the other white dwarf that adds the mass. Given enough time and a close enough binary system, it's possible for the two white dwarfs to gradually fall inward into each other. When they finally collide, the sum of the two masses is more than 1.44 solar masses, triggering the supernova. In this case, there will truly be nothing left behind after the explosion. Astronomers are still working out what fraction of the observed supernovae occur through this double white dwarf pathway. One of the ways we can start to figure this out is by looking at the supernova remnants in our own galaxy where there is no stellar remnant at the core of the explosive bubbles. We've certainly seen this in some cases, such as the supernova recorded in historical records in 1006 CE—even with modern telescopes no remnant has been found. In any case, both pathways are available, and both occur. These detonations, as destructive as they are, have expanded our understanding of the distances in the Universe.

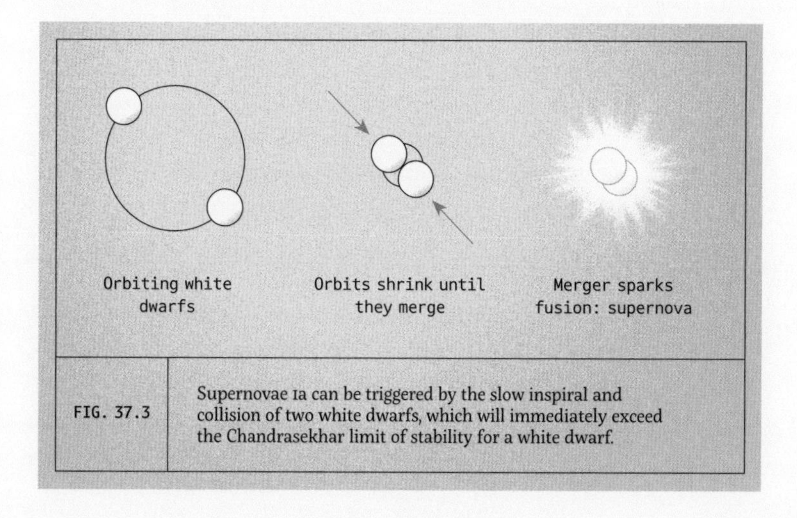

FIG. 37.3 Supernovae Ia can be triggered by the slow inspiral and collision of two white dwarfs, which will immediately exceed the Chandrasekhar limit of stability for a white dwarf.

tracing an expanding Universe

Detecting an unexpected faintness can help us to know both the stars and the Universe they inhabit. Supernovae Ia provide another way of measuring much larger distances than the Cepheid variable stars. In the early 1920s, astronomers had begun to work out the distances to the nearest galaxies, where Cepheid variable stars were detectable. From these initial observations, they noticed that every single galaxy appeared to be redshifted—the light was shifted redward relative to what we'd expect if the galaxies were at rest. This, in turn, meant that they had to be moving away from us. Some galaxies should be expected to be doing this, but every single galaxy moving away from us required some additional thought. And the further away the galaxy was, the faster it was moving away.

A new interpretation was born: the Universe had to be expanding. Rather than having each galaxy flee from our position in the Universe specifically, it's much more straightforward to assume that the galaxies are roughly stationary relative to their local space, but that the space between the galaxies is growing with time. This was already quite a revelation about the Universe we live in, gained simply from looking at Cepheid variable stars within the nearest galaxies, but the story wasn't complete there.

Observations of supernovae Ia allow us to extend this mapping of redshift and measured distance out to even larger values, using the overlap in galaxies with both detectable Cepheid variable stars and supernova explosions. If the Universe were simply expanding at a constant rate, we'd expect to see a continuation of the fixed, straight-line relationship we see in the local Universe no matter how far away we look: the further away the supernova is, as measured by its apparent brightness, the higher its redshift (fig. 38.1).

The distance inferred from the brightness or faintness of the object is known as the luminosity distance, and it is sensitive to what the Universe has been doing since the light from a supernova was released. If the light were traveling in a non-expanding Universe, this would be straightforward—it is just the distance between the two objects that the light has to cross. In a constantly expanding Universe, the light has more distance to travel,

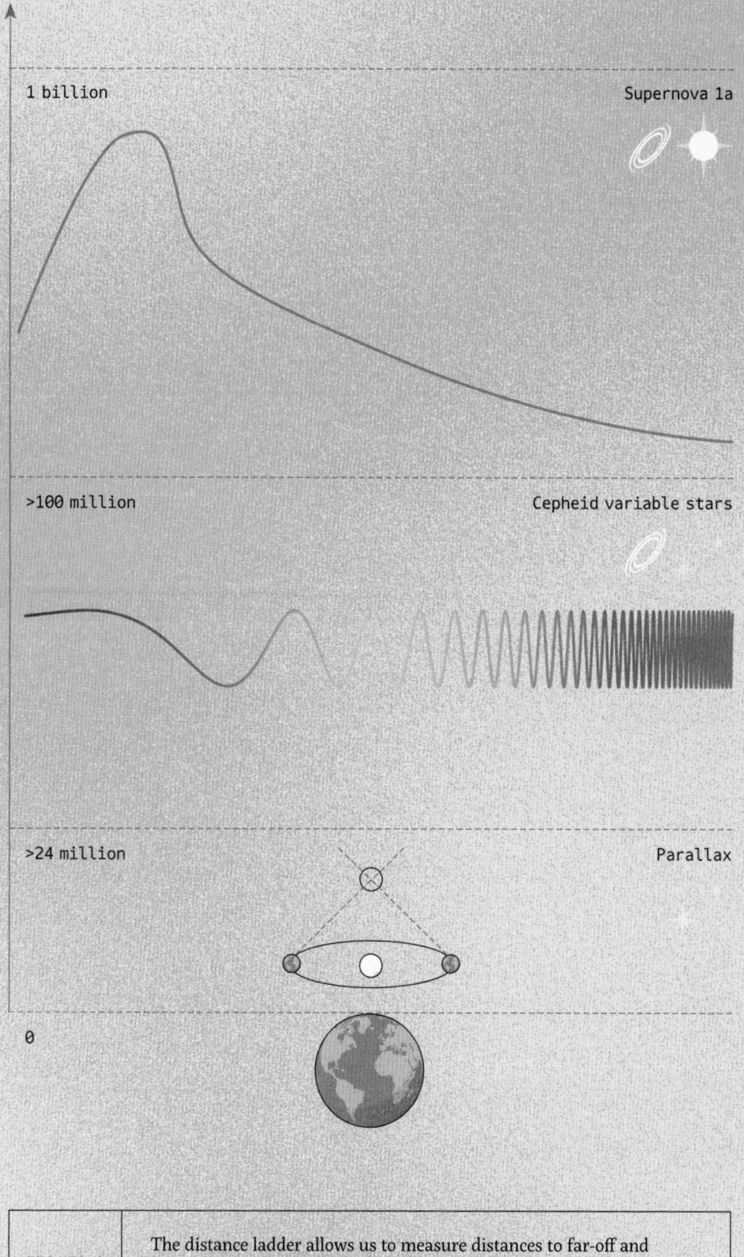

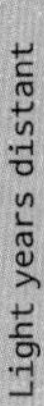

FIG. 38.1	The distance ladder allows us to measure distances to far-off and faint objects, by shifting from parallax, to Cepheid variable stars, to the brightnesses of supernovae.

because the Universe is continuing to expand while the light is traversing the distance. Because the light then travels further, when it reaches us it is fainter than it would be if the Universe were not expanding. If you dig even further into it, if the expansion of the Universe is speeding up or slowing down, then the amount of distance the light has to cover (relative to a fixed expansion) changes slightly. If the Universe's expansion is slowing, then there will be slightly less distance to negotiate, and objects like our supernova should be slightly brighter than we'd predict from a constant expansion model. Conversely, if the expansion is accelerating, then there will be more space to bridge than expected from constant expansion, and so the supernova will appear fainter than predicted.

None of these effects would show up over small distances, so the supernovae that can help distinguish these scenarios are the ones that are the furthest away possible to detect. The more space their light needs to travel, the more the cumulative effect of slightly less or slightly more volume to traverse through will add up. It was in 1998 and 1999 that the first surveys began to find supernovae far enough away to give us the first indications of which scenario was in play (fig. 38.2).

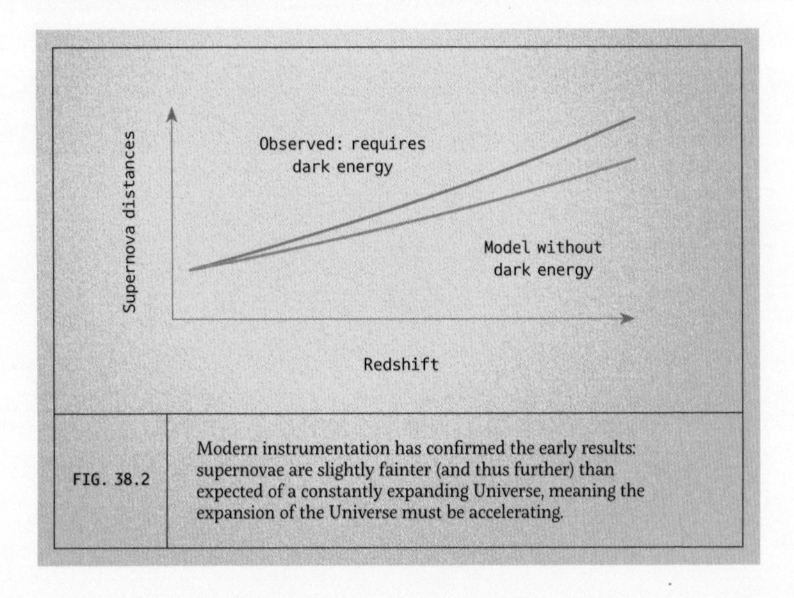

FIG. 38.2 Modern instrumentation has confirmed the early results: supernovae are slightly fainter (and thus further) than expected of a constantly expanding Universe, meaning the expansion of the Universe must be accelerating.

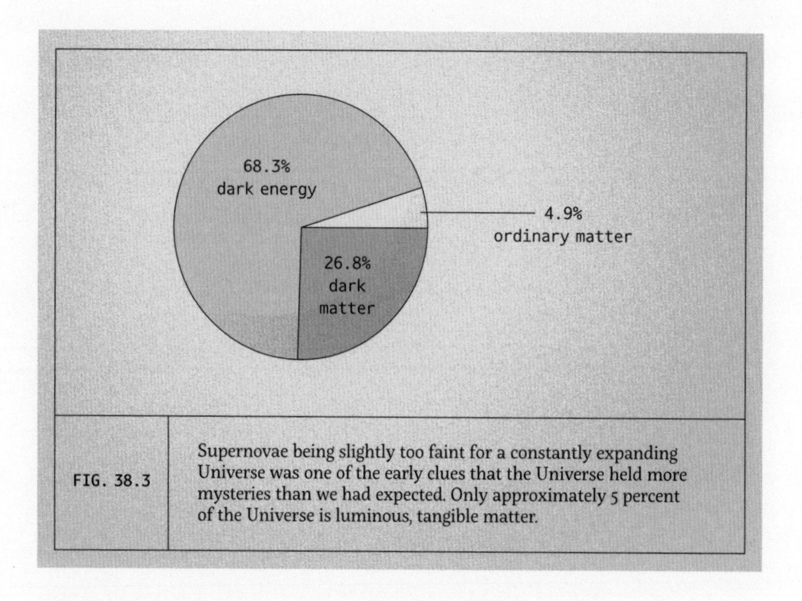

FIG. 38.3 Supernovae being slightly too faint for a constantly expanding Universe was one of the early clues that the Universe held more mysteries than we had expected. Only approximately 5 percent of the Universe is luminous, tangible matter.

Independent teams of scientists found the same thing. The supernovae were slightly too faint to map onto our expectations for a constantly expanding Universe, which meant that the Universe has to be speeding up its expansion.

From here, it got more interesting. The discrepancy between the constant expansion model and what we saw allowed for estimations of how much energy had to be involved in making all of space less dense in this way. From what we understood, gravity ought to be slowly pulling all the galaxies together, and there was no known force that could repel massive objects in a way that could expand the space between all the galaxies. So whatever was responsible for overcoming the gravity of all the galaxies, there had to be a lot of it.

Current models suggest that this source of expansion, which we have conveniently dubbed "dark energy" since we have no clear idea what it is, makes up about 70 percent of the energy in the Universe. Another 25 percent of the mass and energy in our Universe seems to be the dark matter that surrounds every galaxy. In the final 5 percent, we find every single star, grain of dust, atom of gas, complex molecule—all of the luminous matter in our Universe, including the planets, and all the life on Earth (fig. 38.3).

as hints of the early Universe

To know a star, we can look at the impact of the earliest stars on our Universe. While these have been elusive to detect directly, we've made some progress toward that goal. In the meantime, we're starting to understand what these stars must have been like, and the effect they had on the Universe.

One of the reasons we think these stars must have been very massive is the absence of metals in the giant gas cloud from which they formed. The presence of metals helps with the cooling process of a gas cloud. Without metals, it's more likely that a gas cloud will collapse down into one very large star, instead of fragmenting during collapse into a number of smaller stars. The presence of metals therefore not only changes the composition of the stars, but also changes how the stars are able to form in the first place. This, in turn, gives us a natural explanation for why we don't see such massive stars today. There are enough metals in the gas clouds these days that the gas will fragment during its collapse. It's thought that these early stars would have been much more massive than the Sun, with many of them reaching several hundred solar masses, potentially up to 1,000 times the mass of the Sun (fig. 39.1).

There are two places we can look for these first stars. The first is in the extremely early Universe, when it was truly forming its very first stars and putting an end to the Cosmic Dark Ages, when there were no stars yet formed. This first generation of stars has—so far—been extremely hard to spot, in part because the distances the light would have to travel to reach us are so vast that the light from these stars, as bright as they would have been, would now be very faint. As our telescopes become more and more sensitive, it's possible we may be able to see them in the future.

The second place to look is within a galaxy that has already formed some stars, and so is a gravitationally massive object, but which hasn't yet consumed all of its initial reservoir of pristine gas. Computer models of how galaxies form suggest that there may be pockets of gas that aren't affected by the first round of explosions. So the formation of these "first stars" may be prolonged in some galaxies, while the gas that hasn't yet formed into a star gradually

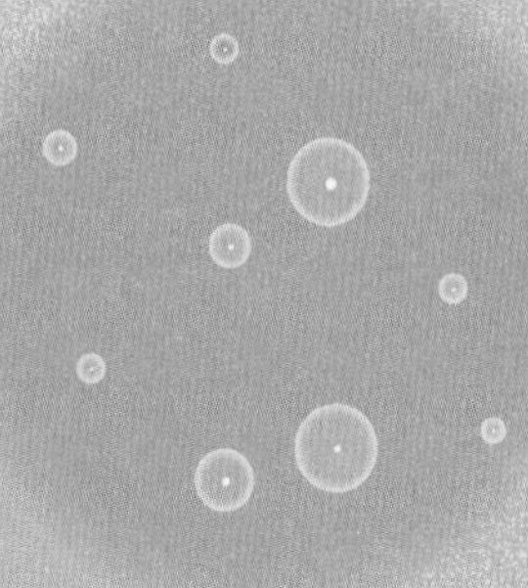

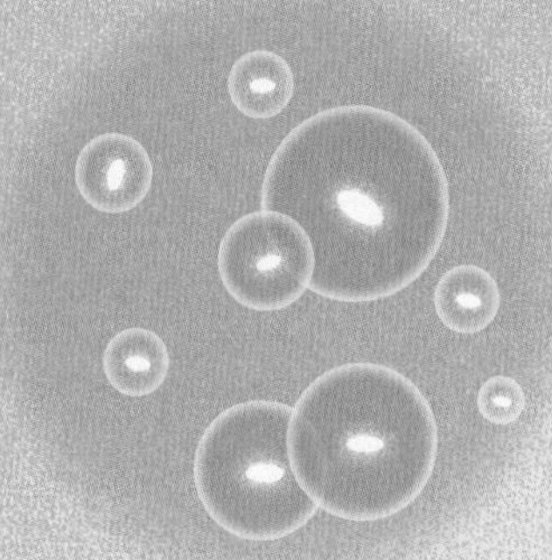

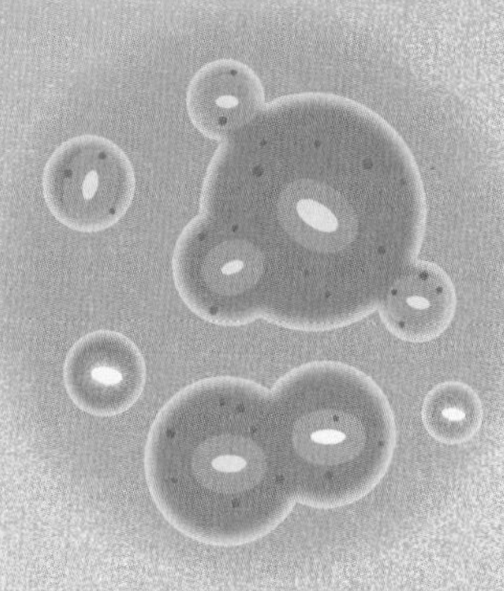

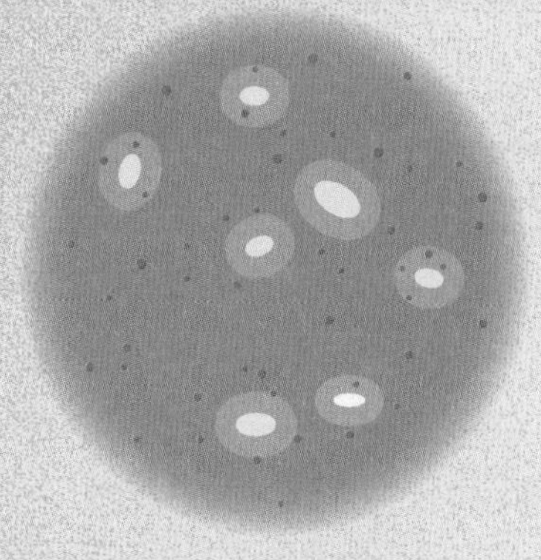

FIG. 39.1 The high-energy light from the earliest stars is thought to be the reason the Universe is now transparent, as it gradually ionized the hydrogen gas between the forming galaxies.

collapses down into a late round of first star formation. These galaxies, while still very distant and hard to observe due to their faintness, are relatively speaking much easier to detect than the true first round of star formation. And it's in this kind of environment that we've done most of our hunting for the signatures of the kinds of stars that would have formed at the earliest times.

In both cases, such massive stars will run at extremely high temperatures, and these temperatures mean that the light they put out into the early Universe would be extremely high energy, with

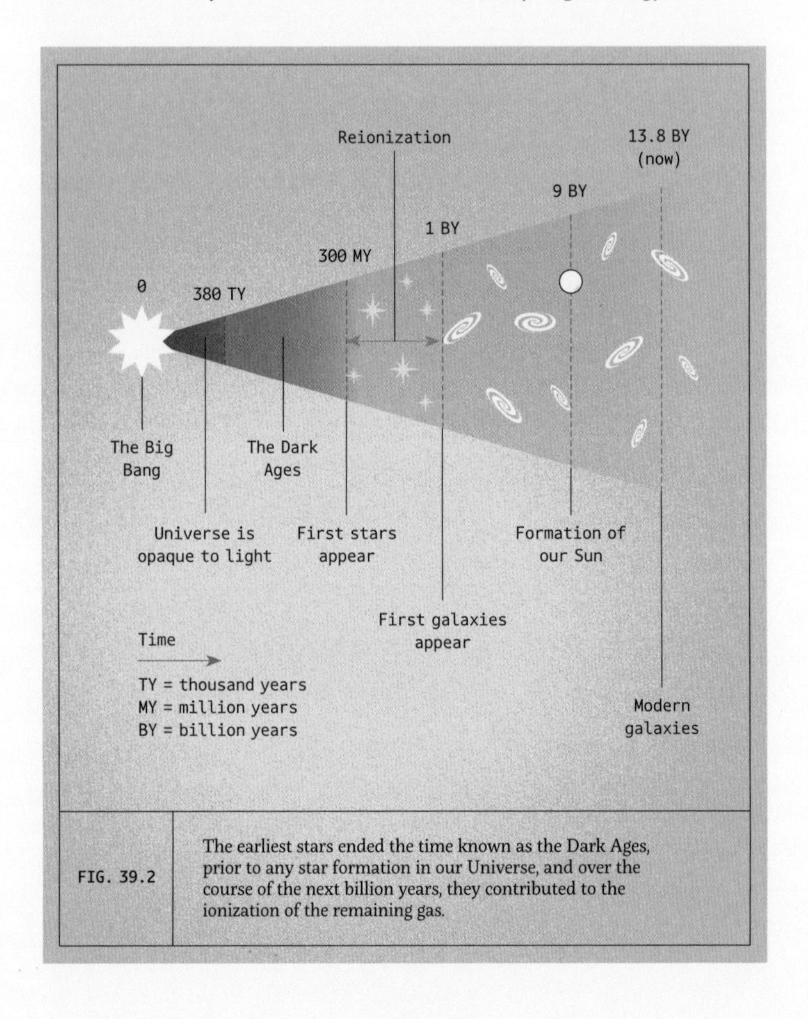

FIG. 39.2 The earliest stars ended the time known as the Dark Ages, prior to any star formation in our Universe, and over the course of the next billion years, they contributed to the ionization of the remaining gas.

a lot of ultraviolet light released. Ultraviolet light is energetic enough to remove an electron from hydrogen, ionizing it. At these temperatures, we'd also expect the starlight to be able to fully ionize helium, removing both of its electrons, which takes more work. With the galaxy still quite full of gas, we can reasonably expect that such massive stars would begin to ionize large bubbles of gas surrounding them, no matter how much helium might be present (fig. 39.2).

One recent study has gone looking for this glow of ionized gas in a well-known ancient galaxy, known as GN-z11, which we looked at as it was only 430 million years after the Big Bang. Researchers were looking for pockets of untouched ionized gas in an otherwise established galaxy, and they found a cloud of what seems to be helium, without any more complex elements within it. If this result holds, they conclude that they're seeing gas ionized by these oldest stars, which must be there, even if invisible.

If this prodigious amount of ionizing light from these oldest stars can escape the galaxy—not an easy feat, considering how much gas there is to run into—then it's likely that these stars, formed at large rates in early galaxies, are also responsible for another facet of the Universe that we might take for granted: its complete transparency. In the very early Universe, the hydrogen that filled the spaces between the dense proto-galaxies was electrically neutral, and neutral hydrogen absorbs certain colors of visible light. It wasn't until this gas could be ionized, which requires the kind of large volume of ultraviolet light produced by the first stars, that all visible light could freely travel through the Universe without being absorbed.

in silver and gold

It is the remains of high mass stars that are responsible for the metals we use in our jewelry. A lot of our jewelry is made of silver, gold, or platinum, and the creation of these metals has a unique pathway.

Precious metals had been a puzzle for a while. While they can be formed during the supernova of a high mass star, it didn't seem like the supernova explosion was all that efficient a producer of silver and gold. Either we hadn't understood the mechanics of supernovae particularly well, or there must be some other process that creates these metals in order for us to see them in the quantities we do on Earth.

The solution to this puzzle pulled together several mysteries at once. The first was an event known as a gamma ray burst. These are unimaginatively named bursts of gamma rays, which come in two durations: short, which last less than two seconds, and long, which carry on emitting gamma rays for longer than two seconds. Long gamma ray bursts are thought to be part of the traditional supernova explosion, but the short gamma ray bursts were perplexing. The best guess, for quite some time, was that they were part of the explosion resulting from the collision of two neutron stars.

Simultaneously, a puzzling explosion dubbed a kilonova, which was not as bright as a supernova but far brighter than a nova, had been observed. By 2013 evidence was mounting that short gamma ray bursts and kilonovae might be part of the same event. And if the short gamma ray burst was coming to us from the collision of two dense astrophysical objects, then so must the kilonova (fig. 40.1).

Definitive evidence that such a cataclysm was the source of our two mysterious outbursts had to wait until 2017, with the detection of gravitational waves, a gamma ray burst, and a kilonova simultaneously. The gravitational wave signal, once interpreted, told us that this specific event involved two dense gravitational objects with masses best explained by two neutron stars—one object was about 2 solar masses, and the other about 1. The less massive object could not have been a white dwarf because the orbits of the two

FIG. 40.1	Kilonovae are dramatic explosions that mark the end of two neutron stars. The explosion releases gamma rays and visible light, and produces a large volume of heavy elements.

objects around each other just before merging were so close that a white dwarf would have occupied more space than was available.

A tremendous number of telescopes were mobilized to investigate the site of the explosion, and the aftermath was rapidly examined across the entire electromagnetic spectrum, trying to learn as much as possible about this event while the opportunity presented itself.

Looking at the elements present in the debris of the explosion, this kilonova was found to produce a huge volume of heavy elements, especially gold, silver, and platinum. In fact, these neutron star mergers were so efficient at creating these heavy elements that neutron stars alone, at the rate they're expected to merge, could account for all the gold and platinum we see on Earth. We wouldn't need to have much—or any—contribution from massive stars undergoing a supernova explosion, and so the

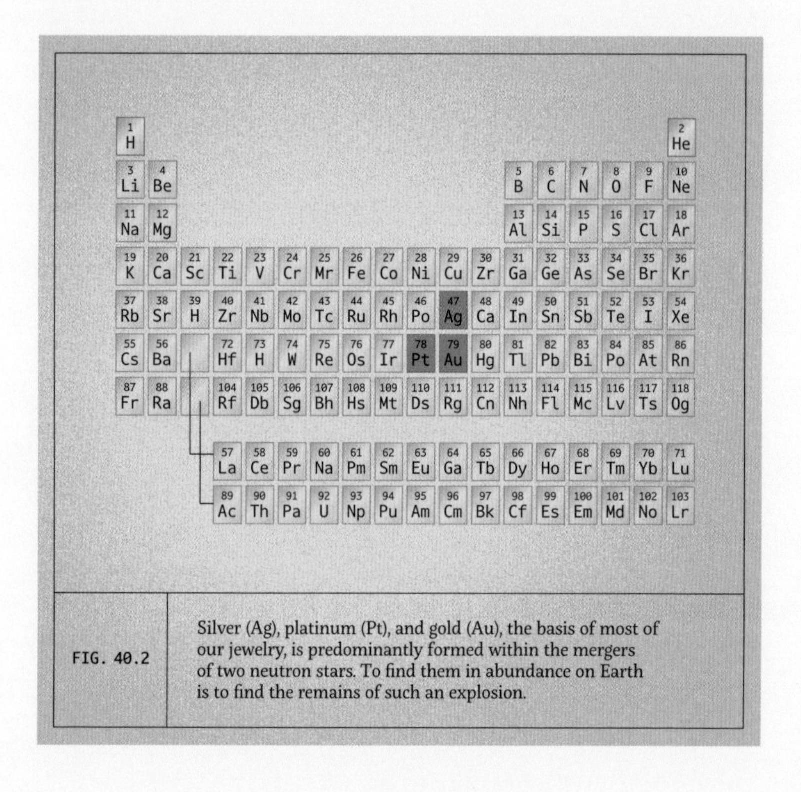

FIG. 40.2 Silver (Ag), platinum (Pt), and gold (Au), the basis of most of our jewelry, is predominantly formed within the mergers of two neutron stars. To find them in abundance on Earth is to find the remains of such an explosion.

inefficiency of the supernova in creating these elements is no longer a problem. They may simply be not the source of the jewelry we wear at all.

Instead, with a single kilonova event able to create more than 15,000 times the Earth's mass in heavy elements, it's much more likely that these collisions between neutron stars are able to rapidly enrich the galaxy they inhabit with heavy elements. Because they are such efficient factories of heavy metals, you don't need too many of them to reproduce the amount of metals we see on Earth (fig. 40.2).

In fact, one study examined the presence of heavy metals in meteorites, formed when our solar system was new. This concluded that the gold and silver in the solar system—along with a large amount of the plutonium—could have been deposited in the gas cloud that would eventually form our system by a single, relatively nearby neutron star merger, perhaps 80 million years before the solar system began to form.

Regardless of whether supernovae also contribute strongly to the presence of the precious metals we have on Earth, it's certain that the collisions between neutron stars are absolutely capable of enriching the regions surrounding them with a haze of gold and platinum. In either case, our jewelry comes to us with an explosive past. The rings and necklaces we create for ourselves were first forged in the destruction of a star and carried outward to mix with gas nearby, which then happened to collapse down into our solar system. There they could be mixed in with the rocky material that formed the Earth.

resources & references

By chapter:

1) www.iau.org/public/themes/constellations • Falchi, F., et al., SciA, 2, 2016, e1600377

2) https://imagine.gsfc.nasa.gov/features/cosmic/nearest_star_info.html • https://chandra.harvard.edu/edu/formal/icecore/The_Historical_Sunspot_Record.pdf • www.nasa.gov/solar-system/nasa-enters-the-solar-atmosphere-for-the-first-time-bringing-new-discoveries/

4) https://nssdc.gsfc.nasa.gov/planetary/factsheet/sunfact.html

5) www.pas.rochester.edu/~emamajek/memo_star_dens.html • https://chandra.harvard.edu/xray_sources/pdf/brown_dwarfs.pdf

6) Ohnaka, K., et al., *A&A*, 2011, 529, A163

7) Raghavan, D., et al., *ApJS*, 2010, 190, 1

8) www.britannica.com/science/magnitude-astronomy • https://nssdc.gsfc.nasa.gov/planetary/factsheet/sunfact.html • https://skyserver.sdss.org/dr1/en/proj/advanced/hr/simplehr.asp

9) https://lco.global/spacebook/stars/protostar/ • https://science.nasa.gov/mission/hubble/science/explore-the-night-sky/hubble-messier-catalog/messier-45/

10) https://spacemath.gsfc.nasa.gov/Grade35/10Page6.pdf • https://science.nasa.gov/jupiter/jupiter-facts/ • https://science.nasa.gov/resource/infographic-profile-of-planet-51-pegasi-b/ • https://news.ucsc.edu/2023/06/helium-tails.html • https://nssdc.gsfc.nasa.gov/planetary/factsheet/uranusfact.html • https://science.nasa.gov/exoplanets/super-earth/ • https://science.nasa.gov/resource/55-cancri-e-skies-sparkle-above-a-never-ending-ocean-of-lava/

11) www.swpc.noaa.gov/phenomena/coronal-mass-ejections • Oughton, E. J., et al., *SpWea*, 2017, 15, 65 • Green, J. L., Boardsen, S., AdSpR, 2006, 38, 130 • Boteler, D. H., *SpWea*, 2019, 17, 1427 • https://science.nasa.gov/science-research/planetary-science/23jul_superstorm/ • https://www.lloyds.com/news-and-insights/risk-reports/library/solar-storm • https://science.nasa.gov/science-news/science-at-nasa/2015/11may_aurorasonmars • https://science.nasa.gov/missions/hubble/hubble-captures-vivid-auroras-in-jupiters-atmosphere/ • www.nasa.gov/image-article/saturns-auroras/

13) Burrows, A., et al., *ApJ*, 1997, 491, 856 • https://chandra.harvard.edu/xray_sources/pdf/brown_dwarfs.pdf • https://science.nasa.gov/solar-system/temperatures-across-our-solar-system/ • www.jpl.nasa.gov/news/coldest-brown-dwarfs-blur-star-planet-lines • https://webbtelescope.org/contents/media/images/4196-Image • www.caltech.edu/about/news/bands-clouds-swirl-across-brown-dwarfs-surface • www.universetoday.com/108693/stormy-with-a-chance-of-molten-iron-rain-first-ever-map-of-exotic-weather-on-brown-dwarfs/#ixzz2rtfsBLYL • www.nasa.gov/missions/spitzer/nasa-helps-decipher-how-some-distant-planets-have-clouds-of-sand • Miles, B. E., et al., *ApJL*, 2023, 946, L6 • Cranmer, S. R., *RNAAS*, 2021, 5, 201

14) http://burro.case.edu/Academics/Astr221/LifeCycle/redgiant.html • www.astronomy.ohio-state.edu/thompson.1847/1101/lecture_evolution_low_mass_stars.html

15) Badenes, C., et al., *ApJL*, 2015, 804, L25 • Lago, P. J. A., et al., *RMxAA*, 2016, 52, 329 • https://science.nasa.gov/mission/hubble/science/explore-the-night-sky/hubble-messier-catalog/messier-57/ • Pottasch, S. R., et al., *A&A*, 2010, 517, A95 • https://hubblesite.org/contents/media/images/2020/31/4682-Image

16) Kepler, S. O., et al., *MNRAS*, 2007, 375, 1315 • https://nssdc.gsfc.nasa.gov/planetary/factsheet/sunfact.html • Provencal, J. L., et al., *ApJ*, 1998, 494, 759 • Mestel, L., *MNRAS*, 1952, 112, 583 • Sackmann, I.-J., et al., *ApJ*, 1993, 418, 457 Kepler, S. O., et al., *MNRAS*, 2007, 375, 1315 https://esahubble.org/images/heic0516c/ • http://vega.lpl.arizona.edu/sirius/A6.html

17) https://astronomy.swin.edu.au/cosmos/C/Classical+Novae • Lloyd, H. M., et al., *MNRAS*, 1997, 284, 137

18) http://hyperphysics.phy-astr.gsu.edu/hbase/Astro/startime.html • Mittag, M., et al., *A&A*, 2023, 669, A9

19) Joyce, M., et al., *ApJ*, 2020, 902, 63 • http://hyperphysics.phy-astr.gsu.edu/hbase/Astro/startime.html • Dolan, M. M., et al., *ApJ*, 2016, 819, 7 • Wheeler, J. C., et al., *A&G*, 2023, 64, 3.11 • Montargès, M., et al., *Nature*, 2021, 594, 365 Dupree, A. K., et al., *ApJ*, 2020, 899, 68 https://hubblesite.org/contents/news-releases/2020/news-2020-44

20) https://astronomy.swin.edu.au/cosmos/c/core-collapse • https://web.archive.org/web/20210420231445/https://websites.pmc.ucsc.edu/~glatz/astr_112/lectures/notes17.pdf • Smartt, S. J., *ARA&A*, 2009, 47, 63 Thielemann, F.-K., et al., *ApJ*, 1996, 460, 408

21) Özel, F., et al., *ApJ*, 2012, 757, 55 • www.jpl.nasa.gov/infographics/neutron-stars • Morales, J. A., et al., *MNRAS*, 2022, 517, 5610 • Berger, A., "Magnetic resonance imaging," BMJ, 2002, 324(7328) • https://science.nasa.gov/missions/webb/webb-finds-evidence-for-neutron-star-at-heart-of-young-supernova-remnant/ • https://nanograv.org/science/topics/pulsars-cosmic-clocks • Smith, D. A., et al., *ApJ*, 2023, 958, 191 • Collins, G. W., et al., *PASP*, 1999, 111, 871 • Zhou, S., et al., *Univ*, 2022, 8, 641

23) Gaia Collaboration, et al., *A&A*, 2023, 674, A1 • Zic, A., et al., *ApJ*, 2020, 905, 23 • Fuhrmeister, B., et al., *A&A*, 2022, 663, A119 • www.aavso.org/vsx/index.php?view=detail.top&oid=9237 • Clementini, G., et al., *A&A*, 2023, 674, A18 Humphreys, R. M., et al., *PASP*, 1994, 106, 1025

24) Mowlavi, N., et al., *A&A*, 2023, 674, A16 • Kirk, B., et al., *AJ*, 2016, 151, 68 • Prša, A., et al., *ApJS*, 2022, 258, 16 • Kolbas, V., et al., *MNRAS*, 2015, 451, 4150 • Richards, M. T., et al., *ApJ*, 2012, 760, 8 • Lucy, L. B., *ApJ*, 1976, 205, 208 • Almeida, L. A., et al., *ApJ*, 2015, 812, 102

25) Bland-Hawthorn, J., et al., *ARA&A*, 2016, 54, 529 • Blitz, L., et al., *ApJ*, 1991, 379, 631 • Poggio, E., et al., *NatAs*, 2020, 4, 590 • Ramos, P., et al., *A&A*, 2022, 666, A64 • Lynden-Bell, D., et al., *MNRAS*, 1995, 275, 429

26) https://science.nasa.gov/sun/facts • https://public.nrao.edu/ask/what-causes-the-suns-periodic-vertical-oscillation-through-the-plane-of-the-galaxy/

27) https://science.nasa.gov/missions/hubble/hubble-views-the-star-that-changed-the-universe/ • Hubble, E. P., *ApJ*, 1926, 64, 321

28) Kroupa, P., et al., *MNRAS*, 2001, 321, 699 • Alvarez-Baena, N., et al., *A&A*, 2024, 687, A101

29) Jurić, M., et al., *ApJ*, 2008, 673, 864 • https://astronomy.swin.edu.au/cosmos/T/Thick+Disk • https://astronomy.swin.edu.au/cosmos/B/Bulges • Helmi, A., *A&ARv*, 2008, 15, 145 • https://astronomy.swin.edu.au/cosmos/S/Stellar+Halo

30) Rubin, V. C., et al., *AJ*, 1962, 67, 491 • Corbelli, E., et al., *MNRAS*, 2000, 311, 441 • Sharma, G., et al., *A&A*, 2021, 653, A20 • Lovell, M. R., et al., *MNRAS*, 2018, 481, 1950

31) Tolstoy, E., et al., *ARA&A*, 2009, 47, 371 • https://earthsky.org/clusters-nebulae-galaxies/what-is-the-local-group/

33) GRAVITY Collaboration, et al., *A&A*, 2023, 677, L10 • www.eso.org/public/images/eso2208-eht-mwe/ • https://eventhorizontelescope.org/blog/astronomers-reveal-first-image-black-hole-heart-our-galaxy • GRAVITY Collaboration, et al., *A&A*, 2018, 615, L15 • Häberle, M., et al., *Nature*, 2024, 631, 285

34) Leavitt, H. S., *AnHar*, 1907, 60, 87 • Leavitt, H. S., et al., *HarCi*, 1912, 173 • https://science.nasa.gov/mission/hipparcos/ • https://skyserver.sdss.org/dr1/en/proj/advanced/hr/hipparcos1.asp • Gaia Collaboration, et al., *A&A*, 2023, 674, A1 • https://science.nasa.gov/missions/hubble/hubble-views-the-star-that-changed-the-universe/

35) Xiang, M., et al., *Nature*, 2022, 603, 599 • Förster Schreiber, N. M., et al., *ARA&A*, 2020, 58, 661 • Wright, E. L., *PASP*, 2006, 118, 1711

37) Hillebrandt, W., et al., *ARA&A*, 2000, 38, 191 • González Hernández, J. I., et al., *Nature*, 2012, 489, 533 • Phillips, M. M., *ApJL*, 1993, 413, L105

38) Riess, A. G., et al., *AJ*, 1998, 116, 1009 • Perlmutter, S., et al., *ApJ*, 1999, 517, 565

39) https://science.nasa.gov/missions/webb/webb-unlocks-secrets-of-one-of-the-most-distant-galaxies-ever-seen/ • Maiolino, R., et al., *A&A*, 2024, 687, A67 • Naidu, R. P., et al., *ApJ*, 2020, 892, 109

40) www.nasa.gov/news-release/nasas-hubble-finds-telltale-fireball-after-gamma-ray-burst/ • Abbott, B. P., et al., *ApJL*, 2017, 848, L12 • Pian, E., et al., *Nature*, 2017, 551, 67 • Kasen, D., et al., *Nature*, 2017, 551, 80 Rastinejad, J. C., et al., *Nature*, 2022, 612, 223

index

acknowledgments

From the author
I'd like to thank everyone who made this book possible. In particular, my thanks to my agent, Peter Tallack; editor Duncan Heath; project manager, Blanche Craig; and designer, Lindsey Johns, along with Slav Todorov and Jason Hook of UniPress. Thanks also to the friends and family who continuously encourage me to go for it.

Finally, my gratitude to the community of astrophysicists who not only work extremely hard to uncover new patterns and meanings behind our observations of the cosmos, but make that knowledge available. Thank you for your papers.

Image credits
Cover ESA/Hubble & NASA; **2** ESA/Webb, NASA & CSA, A. Scholz, K. Muzic, A. Langeveld, R. Jayawardhana; **9** Shutterstock/shooarts; **14** NASA; **25** NASA/GRC/Jordan Salkin; **35** ESA/Hubble and NASA; **49** NASA, ESA, CSA, STScI/Joseph DePasquale (STScI), Alyssa Pagan (STScI), Anton M. Koekemoer (STScI); **75** NASA, ESA, C. R. O'Dell (Vanderbilt University), and D. Thompson (Large Binocular Telescope Observatory); **89** NASA, ESA, CSA, STScI, Webb ERO Production Team; **103** NASA, ESA, J. Hester, and A. Loll (Arizona State University); **105** Shutterstock/Kovalto1; **110**, **114** Wikimedia Commons; **117** NASA, ESA, N. Smith (University of Arizona), and J. Morse (BoldlyGo Institute); **119** Wikimedia Commons; ESA/Gaia/DPAC, CC BY-SA 3.0 IGO; **142**, **145** Shutterstock/RNko7; **149** ESA/Hubble & NASA; **155** Wikimedia Commons; **160** ESA/ATG medialab; **167** Shutterstock/RNko7; **171** NASA, ESA, the Hubble Heritage Team (STScI/AURA)-ESA/Hubble Collaboration, and A. Evans (University of Virginia, Charlottesville/NRAO/Stony Brook University).

Published by
Princeton Architectural Press
A division of Chronicle Books LLC
70 West 36th Street
New York, NY 10018
papress.com

Printed and bound in China
28 27 26 25 4 3 2 1 First edition

ISBN 978-1-7972-3500-4
Library of Congress Control Number: 2024948049

Typeset in Servus Slab and Darkmode Mono Off

This book was conceived, designed, and produced by UniPress Books Limited

Publisher: Jason Hook
Managing editor: Slav Todorov
Art director: Alex Coco
Project manager: Blanche Craig
Design and illustration: Lindsey Johns